Hardware Accelerators in Data Centers

Christoforos Kachris · Babak Falsafi
Dimitrios Soudris
Editors

Hardware Accelerators in Data Centers

Springer

Editors
Christoforos Kachris
Microprocessors and Digital Systems Lab
National Technical University of Athens
Athens, Greece

Dimitrios Soudris
National Technical University of Athens
Athens, Greece

Babak Falsafi
IC IINFCOM PARSA
École Polytechnique Fédérale de Lausanne
Lausanne, Switzerland

ISBN 978-3-030-06518-8 ISBN 978-3-319-92792-3 (eBook)
https://doi.org/10.1007/978-3-319-92792-3

Softcover re-print of the Hardcover 1st edition 2019

This Springer imprint is published by the registered company Springer Nature Switzerland AG
The registered company address is: Gewerbestrasse 11, 6330 Cham, Switzerland

Preface

Emerging cloud applications like machine learning, artificial intelligence (AI), deep neural networks, and big data analytics have created the need for more powerful data centers that can process huge amounts of data without consuming excessive amounts of power. To face these challenges, data center operators have to adopt novel architectures with specialized high-performance energy-efficient computing systems, such as hardware accelerators, that can process the increasing amount of data in a more energy-efficient way. To this end, a new era has emerged; the era of heterogeneous distributed computing systems that consist of contemporary general-purpose processors (CPUs), general-purpose graphic processing units (GP-GPUs), and field-programmable gate arrays (FPGAs or ACAP Adaptive Compute Acceleration Platform). The utilization of several heterogeneous architectures poses several challenges in the domain of efficient resource utilization, efficient scheduling, and resource management.

In this book, we have collected the most promising and the most recent research activities in this emerging domain of heterogeneous computing that is based on the efficient utilization of hardware accelerators (i.e., FPGAs). The book contains 13 system-level architectures that show how to efficiently utilize hardware accelerators in the data centers to face emerging cloud applications. The proposed architectures tackle several challenges such as the energy efficiency of hardware accelerators in data centers, the resource utilization and management, and the programmability of heterogeneous infrastructure based on the hardware accelerators. We hope this book serves as a reference book for any researcher, engineer, and academic works on the exciting new area of heterogeneous computing with accelerators.

Athens, Greece
April 2018

Christoforos Kachris

Contents

Chapter 1
Introduction

Christoforos Kachris, Babak Falsafi and Dimitrios Soudris

1.1 Introduction

Emerging applications like cloud computing, machine learning, AI and big data analytics require powerful systems that can process large amounts of data without consuming high power. Furthermore, these emerging applications require fast time-to-market and reduced development times. To address the large processing requirements of emerging applications, novel architectures are required to be adopted by the data center vendors and the cloud computing providers.

Relying on Moore's law, CPU technologies have scaled in recent years through packing an increasing number of transistors on chip, leading to higher performance. However, on-chip clock frequencies were unable to follow this upward trend due to strict power-budget constraints. Thus, a few years ago, a paradigm shift to multicore processors was adopted as an alternative solution for overcoming the problem. With multicore processors, we could increase server performance without increasing their clock frequency. Unfortunately, this solution was also found not to scale well in the longer term. The performance gains achieved by adding more cores inside a CPU come at the cost of various, rapidly scaling complexities: inter-core communication, memory coherency and, most importantly, power consumption [1].

In the early technology nodes, going from one node to the next allowed for a nearly doubling of the transistor frequency, and, by reducing the voltage, power density remained nearly constant. With the end of Dennard's scaling, going from

C. Kachris (✉)
Institute of Communication and Computer Systems (ICCS/NTUA),
Athens, Greece
e-mail: kachris@microlab.ntua.gr

B. Falsafi
École Polytechnique Fédérale de Lausanne (EPFL), Lausanne, Switzerland

D. Soudris
Department of Electrical and Computer Engineering,
National Technical University of Athens, Athens, Greece

C. Kachris et al. (eds.), *Hardware Accelerators in Data Centers*,
https://doi.org/10.1007/978-3-319-92792-3_1

one node to the next still increases the density of transistors, but their maximum frequency is roughly the same and the voltage does not decrease accordingly. As a result, the power density increases now with every new technology node. The biggest challenge therefore now consists of reducing power consumption and energy dissipation per mm^2.

Therefore, the failure of Dennard's scaling, to which the shift to multicore chips is partially a response, may soon limit multicore scaling just as single-core scaling has been curtailed [2]. This issue has been identified in the literature as the *dark silicon* era in which some of the areas in the chip are kept powered down in order to comply with thermal constraints [3]. One way to address this problem is through the utilization of hardware accelerators. Hardware accelerators can be used to offload the processor, increase the total throughput and reduce the energy consumption.

1.2 The Era of Accelerators in the Data Centers

As the requirements for processing power of the data centers continue to increase rapidly, higher performance computing systems are required to sustain the increased communication bandwidth demand within the data center. Current server processors cannot affordable satisfy the required computational demand of emerging application without consuming excessive power. Hardware accelerators provide a viable solution offering high throughput, reduced latency and higher energy efficiency compared to current servers based on commodity processors.

This book presents the most recent and promising solutions that have been presented in the area of the computing hardware accelerators in heterogeneous data centers. Hardware accelerator for data centers is an interdisciplinary topic for all the communities that are active in the domain of computer architectures, high-performance computing, data center design and cloud computing.

Hardware accelerators are mainly used in embedded systems to offload the processors for several tasks like compression, encryption, etc. and to provide higher performance and lower energy consumption. Current data centers need to embrace new technologies in order to face the increased network traffic due to emerging applications like cloud computing. Hardware accelerators based on reconfigurable logic (i.e. FPGAs) can provide higher throughput and better energy efficiency.

Hardware accelerator in data centers is an emerging topic that has recently gained attention by major data center and cloud computing vendors. The recent purchase of Altera (a major vendor of FPGA-based accelerators) by Intel, and the collaboration between IBM and Xilinx shows that soon the hardware accelerators will penetrate the hyperscale data center in order to provide higher performance and higher energy efficiency. This book aims to provide an overview of the architectures, programming frameworks and hardware accelerators for typical cloud computing applications in data centers.

The following section gives an overview of the main chapters that are presented in this book.

1.3 Book Chapters

Chapter 2 presents an integrated infrastructure required to support the deployment of FPGAs at a large scale in a data center, developed by University of Toronto. This platform considers FPGAs to be peers to the CPU-based servers rather than using them as accelerator slaves, which is the more common view. The goal is to enable the allocation and use of the FPGAs as computing resources in the same way that current servers are provisioned in a data centre. Their approach is to build on the existing knowledge and experience with provisioning software-based processing elements in the data centre and adapting the current infrastructure to the differences that FPGAs bring. This incurs minimal disruption and adjustment to how systems are currently being deployed.

Chapter 3 presents the dReDBox (disaggregated Recursive Data center in a Box) project. dReDBox addresses the problem of fixed resource proportionality in next-generation, low-power data centers by proposing a paradigm shift towards finer resource allocation granularity, where the unit is the function block rather than the mainboard tray. This introduces various challenges at the system design level, requiring elastic hardware architectures, efficient software support and management, and programmable interconnect. Memory and hardware accelerators can be dynamically assigned to processing units to boost application performance, while high-speed, low-latency electrical and optical interconnect is a prerequisite for realizing the concept of data center disaggregation. This chapter presents the dReDBox hardware architecture and discusses design aspects of the software infrastructure for resource allocation and management. Furthermore, initial simulation and evaluation results for accessing remote, disaggregated memory are presented, employing benchmarks from the Splash-3 and the CloudSuite benchmark suites.

Chapter 4 presents the OPERA projects that aims bring innovative solutions to increase the energy efficiency of cloud infrastructures, by leveraging on modular, high-density, heterogeneous and low-power computing systems, spanning data center servers and remote CPS. The seamless integration of Cyber-Physical Systems (CPS) and cloud infrastructures allows the effective processing of the huge amount of data collected by smart embedded systems, towards the creation of new services for the end users. However, trying to continuously increase data center capabilities comes at the cost of an increased energy consumption. The effectiveness of the proposed solutions in OPERA is demonstrated with key scenarios: a road traffic monitoring application, the deployment of a virtual desktop infrastructure and the deployment of a compact data center on a truck.

Chapter 5 presents a novel framework, called SPynq, for the efficient mapping and acceleration of Spark applications on heterogeneous systems. The Spynq framework has been demonstrated in all-programmable MPSoC-based platforms, such as Zynq and heterogeneous systems with FPGAs attached to PCIe interfaces. Spark has been mapped to the Xilinx Pynq platform, and the proposed framework allows the seamless utilization of the programmable logic for the hardware acceleration of computational intensive machine learning kernels. The required libraries have also been developed that hide the accelerator's details to minimize the design effort to utilize the accelerators.

A cluster of four worker nodes based on the all-programmable MPSoCs has been implemented, and the proposed platform is evaluated in two typical machine learning applications based on logistic regression and k-means. Both the logistic regression and the k-means kernels have been developed as accelerators and incorporated to the Spark. The developed system is compared to a high-performance Xeon server that is typically used in cloud computing. The performance evaluation shows that the heterogeneous accelerator-based MPSoC can achieve up to 2.5× system speedup compared with a Xeon system and 23× better energy efficiency. For embedded application, the proposed system can achieve up to 36× speedup compared to the software only implementation on low-power embedded processors and 29× lower energy consumption.

Chapter 6 presents the Modular Microserver Data centre (M2DC) project that targets the development of a new class of energy-efficient TCO-optimized appliances with built-in efficiency and dependability enhancements. The appliances are easy to integrate with a broad ecosystem of management software and fully software defined to enable optimization for a variety of future demanding applications in a cost-effective way. The highly flexible M2DC server platform enables customization and smooth adaptation to various types of applications, while advanced management strategies and system efficiency enhancements (SEE) are used to improve energy efficiency, performance, security and reliability. Data center capable abstraction of the underlying heterogeneity of the server is provided by an OpenStack-based middleware.

Chapter 7 presents the TANGO project (Transparent heterogeneous hardware Architecture deployment for eNergy Gain in Operation (TANGO)). TANGO project's goal is to characterize factors which affect power consumption in software development and operation for Heterogeneous Parallel Hardware (HPA) environments. Its main contribution is the combination of requirements engineering and design modeling for self-adaptive software systems, with power consumption awareness in relation to these environments. The energy efficiency and application quality factors are integrated in the application lifecycle (design, implementation and operation). To support this, the key novelty of the project is a reference architecture and its implementation. Moreover, a programming model with built-in support for various hardware architectures including heterogeneous clusters, heterogeneous chips and programmable logic devices is provided. This leads to a new cross-layer programming approach for heterogeneous parallel hardware architectures featuring software and hardware modelings.

Chapter 8 investigates how programmable resources can be integrated into the cloud and edge infrastructure. In order to achieve this goal, the programmable cloud platform has been designed and implemented. This is a hardware/software platform that allows for the seamless integration of a programmable logic device into the cloud and edge infrastructure and the deployment of tasks on it over OpenStack, the de facto open-source software for cloud environments. The presented system is based on an FPGA SoC, a device that combines both CPUs and programmable fabric, thus allowing us to optimize the division of labor within the device. The programmable cloud platform includes both software and hardware to allow an FGPA SoC to be integrated seamlessly with the OpenStack worker and controller, and provides an environment where hardware-based user tasks can be deployed. For the evaluation, an h264 decoder is used as an HW task to evaluate the performance of our system in two ways. First, to verify that the SW stack does not impede the HW by causing too much overhead and second, to indeed show that the FPGA can accelerate a common task such as video decoding while been house in a relatively cheap, modest in size programmable device.

Chapter 9 presents the cloud lighting architecture that aims to use of techniques to address the problems emerging from the confluence of issues in the emerging cloud: rising complexity and energy costs, problems of management and efficiency of use, the need to efficiently deploy services to a growing community of non-specialist users and the need to facilitate solutions based on heterogeneous components. This approach attempts to address several issues like energy efficiency, improved accessibility to cloud and support of heterogeneity in the data centers. The CloudLightning architecture is a hierarchical organization of physical infrastructure but unlike traditional organizations it makes use of a resource management framework that is logically hierarchical. The bottom layer of this framework hierarchy consists of many resource managers. These managers are autonomous and, in contrast to traditional systems, each manages a relatively small number of physical resources.

Chapter 10 presents the TULIPP platform for the efficient utilization of FPGA resources in the cloud. TULIPP consists of a hardware system, supportive development utilities and a real-time operating system (RTOS). Chained platforms provide scalability and higher processing power. The project develops and provides a reference hardware architecture—a scalable low-power board, a low-power operating system and image processing libraries and a productivity-enhancing toolchain. The project is use-case driven, providing real-time low-power demonstrators of a medical image processing application, automotive-embedded systems for driver assistance (ADAS) and applications for Unmanned Aerial Vehicles (UAVs). The close connection to its setup ecosystem and standardization organizations will allow the TULIPP project to propose new standards derived from its reference platform and handbook to the industry.

Chapter 11 presents the ECOSCALE architecture. ECOSCALE proposes a scalable programming environment and architecture, aiming to substantially reduce energy consumption as well as data traffic and latency in data centers and HPC. Furthermore, ECOSCALE introduces a novel heterogeneous energy-efficient hierarchical architecture, as well as a hybrid many-core+OpenCL programming environment and runtime system. The ECOSCALE approach is hierarchical and is expected to scale well by partitioning the physical system into multiple independent Workers. Workers are interconnected in a tree-like fashion and define a contiguous global address space that can be viewed either as a set of partitions in a Partitioned Global Address Space (PGAS) or as a set of nodes hierarchically interconnected via an MPI-like protocol. To further increase energy efficiency, as well as to provide resilience, the workers employ reconfigurable accelerators mapped onto the virtual address space by utilizing a dual-stage system memory management unit with coherent memory access. The architecture supports shared partitioned reconfigurable resources accessed by any worker in a PGAS partition, as well as automated hardware synthesis of these resources from an OpenCL-based programming model.

Chapter 12 presents the DOLPHIN project that proposes an approach towards optimizing the energy consumption of federated data centers by means of continuous resources monitoring and by exploiting flexible SLA renegotiation, coupled with the operation of predictive, multilayered optimization components. Such energy consumption management should be subject to limitations of variable elasticity such as the balancing needs of the smart grid, the service-level agreements (SLAs) signed between the cloud services providers and their customers, the actual computational capacity of the data centers and the needs of the data center owners for maximizing profit through service provisioning.

Chapter 13 exploits the increased variability within CPUs and memories manufactured in advanced nanometer nodes that give rise to another type of heterogeneity, the intrinsic hardware heterogeneity which differs from the functional heterogeneity, which is discussed in the previous chapters. In particular, the aggressive miniaturization of transistors led to worsening of the static and temporal variations of transistor parameters, resulting eventually to large variations in the performance and energy efficiency of the manufactured chips. Such increased variability causes otherwise-identical nanoscale circuits to exhibit different performance or power consumption behaviors, even though they are designed using the same processes and architectures and manufactured using the same exact production lines. The UniServer approach discussed in Chap. 13 attempts to quantify the intrinsic variability within the CPUs and memories of commodity servers, and reveals the true capabilities of each core and memory through unique automated online and offline characterization processes. The revealed capabilities and new operating points or cores and memories that may differ substantially from the ones currently adopted by manufactures are then being exploited by an enhanced error-resilient software stack for improving the energy efficiency, while maintaining high levels of system availability. The UniServer approach introduces innovations across all layers of the hardware and system software stacks; from firmware to hypervisor, up to the OpenStack resource manager targeting deployments at the emerging edge or classical cloud data centres.

References

1. Esmaeilzadeh H, Blem E, Amant RS, Sankaralingam K, Burger D (2013) Power challenges may end the multicore era. Commun ACM 56(2):93–102
2. Martin C (2014) Post-dennard scaling and the final years of Moores Law. Technical report
3. Esmaeilzadeh H, Blem E, Amant RS, Sankaralingam K, Burger D (2012) Dark silicon and the end of multicore scaling. IEEE Micro 32(3):122–134
4. Kachris C, Soudris D (2016) A survey on reconfigurable accelerators for cloud computing. In: 2016 26th International conference on field programmable logic and applications (FPL), pp 1–10, Aug 2016

5. Xilinx reconfigurable acceleration stack targets machine learning, data analytics and video streaming. Technical report (2016)
6. Byma S, Steffan JG, Bannazadeh H, Leon-Garcia A, Chow P (2014) FPGAs in the cloud: booting virtualized hardware accelerators with openstack. In: 2014 IEEE 22nd annual international symposium on field-programmable custom computing machines (FCCM), pp 109–116, May 2014
7. Cong J, Huang M, Wu D, Hao Yu C (2016) Invited—heterogeneous datacenters: options and opportunities. In: Proceedings of the 53rd annual design automation conference, DAC'16. ACM, New York, NY, USA, pp 16:1–16:6
8. Apache, spark. http://spark.apache.org/, http://spark.apache.org/
9. Zaharia M, Chowdhury M, Das T, Dave A, Ma J, McCauley M, Franklin MJ, Shenker S, Stoica I (2012) Resilient distributed datasets: a fault-tolerant abstraction for in-memory cluster computing. In: Proceedings of the 9th USENIX conference on networked systems design and implementation, NSDI'12. USENIX Association, Berkeley, CA, USA, pp 2–2
10. Pynq: Pyhton productivity for Zynq. Technical report (2016)

Chapter 2
Building the Infrastructure for Deploying FPGAs in the Cloud

Naif Tarafdar, Thomas Lin, Daniel Ly-Ma, Daniel Rozhko, Alberto Leon-Garcia and Paul Chow

2.1 Introduction

Two of the most significant challenges with cloud computing are performance and power. While performance is always a consideration, the scale of data centres makes power a particularly significant factor. To address these challenges, the computing world has begun to use *accelerators*, principally GPUs to this point. The most common configuration for an accelerator is as a *slave* to a host processor, meaning that the host processor controls how the accelerator is used. In our approach, we consider all computing devices to be peers, which means that there is no requirement for any host to control an accelerator as a *slave*. Any computing device can interact with any other computing device on an equal basis.

In a heterogeneous computing world, starting with the peer model makes it easier to work with well-known and existing parallel programming models that were mostly developed when all processors were homogeneous. We feel that it is important to introduce heterogeneity starting from a point that is familiar to application developers

N. Tarafdar · T. Lin · D. Ly-Ma · D. Rozhko · A. Leon-Garcia · P. Chow (✉)
Department of Electrical and Computer Engineering, University of Toronto,
Toronto, Canada
e-mail: pc@eecg.toronto.edu

N. Tarafdar
e-mail: naif.tarafdar@mail.utoronto.ca

T. Lin
e-mail: t.lin@mail.utoronto.ca

D. Ly-Ma
e-mail: d.lyma@mail.utoronto.ca

D. Rozhko
e-mail: daniel.rozhko@mail.utoronto.ca

A. Leon-Garcia
e-mail: alberto.leongarcia@utoronto.ca

C. Kachris et al. (eds.), *Hardware Accelerators in Data Centers*,
https://doi.org/10.1007/978-3-319-92792-3_2

today. We do this by decoupling the application development from the absolute need to know what type of device architecture will be used to do the computation. With this philosophy, programmers can continue to work with the abstraction that there is a pool of processors and even do their development in the usual way on a traditional cluster of processors. Once their applications are functionally correct running on the software cluster, they, or some future tool, can map some or all of the parallel tasks to appropriate architectures. This has the additional advantages that it removes a layer of complexity from the programmer, i.e. not having to think about what accelerator to use and how to incorporate it into the application, and also provides *portability* because the appropriate architecture may change depending on the requirements and over time. Code portability is extremely important and is what has enabled the development of a vast amount of software code that can be easily reused and adapted. It is important not to lose this property as we move to a more heterogeneous world.

At the University of Toronto, the high-performance reconfigurable computing group focuses its work on making Field-Programmable Gate Arrays (FPGAs) easier to use as computing devices. Our early work of adapting the Message-Passing Interface (MPI) [13] to a heterogeneous platform [21] made it possible for a biochemistry Ph.D. student to help develop a heterogeneous computing platform to do molecular dynamics simulations comprising Xeon processors and FPGAs [15]. By using MPI, the application could be first developed on a standard Linux cluster using a standard MPI library distribution with all the advantages of the software debugging and performance analysis tools available. A later step mapped appropriate computations onto FPGAs that were enabled to use MPI communications. None of the parts that remained in software had to be changed. The success of the MPI work has motivated us to continue the philosophy of working with existing programming models and frameworks and finding ways to adapt them to use FPGAs. This paper describes the achievements and current work we are doing towards building the many layers required to deploy FPGAs in the cloud.

In Sect. 2.2, we describe the abstraction layers used in the software world and we propose an equivalent set of layers for hardware implemented in FPGAs. We then show the additional layers required for deploying applications in the cloud. The abstraction layers help to identify the different functional requirements needed to build a fully heterogeneous computing environment comprising processors and FPGAs. Section 2.3 describes the basic components used within our cloud platform and Sect. 2.4 presents our middleware platform for building heterogeneous FPGA network clusters in our cloud. A slightly different middleware platform for building heterogeneous virtualized network function service chains is described in Sect. 2.5. Work towards supporting multiple tenants sharing an FPGA is described in Sect. 2.6 and an approach to enable the live migration of FPGA computations is described in Sect. 2.7. We end with some final remarks in Sect. 2.8.

2.2 Abstraction Layers

It has always been much more difficult to build computing applications on FPGAs because there is very little infrastructure that supports the applications [9], particularly when it comes to input/output and the memory system. It is common for an FPGA application developer to also be burdened with building the PCIe interface for connection to the host, the driver for the PCIe interface and the memory controller to access the off-chip memory. This must be done again when a new FPGA board is to be used. In the software world, this problem rarely occurs because many abstractions have been developed that software relies on to achieve ease of development and portability. For example, Linux [14] can be considered an abstraction layer that runs on many platforms, so any application built to run on Linux will generally be able to run on any Linux platform.

Figure 2.1 shows the abstraction layers between a software (SW) application and the hardware platform [9], which is typically the system *Motherboard*. Using an MPI application as an example, portability is achieved because the *MPI Program* uses a standard *MPI Library* for message passing. The library is available for several Operating Systems (OS), such as Linux, so the application can be run on different systems as long as there is a library available. The lower levels of the SW OS must adapt to the hardware platform. The vendor of the processor hardware usually provides a thin *firmware* layer call the *BIOS* that can do operations like hardware system checks, initialization and boot the operating system.

Figure 2.2 shows how the software abstraction layers of Fig. 2.1 can be augmented with equivalent hardware abstraction layers [9]. We propose these layers as a way to think about what is required to properly support computing on an FPGA platform. This follows our philosophy of learning from the success of the environments developed for building software applications. The interconnect is any physical link that allows software running on the processor to communicate with the application running on the FPGA. The Board Support Package (BSP) provides facilities similar to a BIOS. In Fig. 2.2, we show the BSP partitioned into a software part (BSPS) and a hardware part (BSPH) because there will typically be some control software that can interact with the BSPS and the BSPH will provide an interface to the components running in the *FPGA Fabric*. The *HW OS* represents a *hardware*

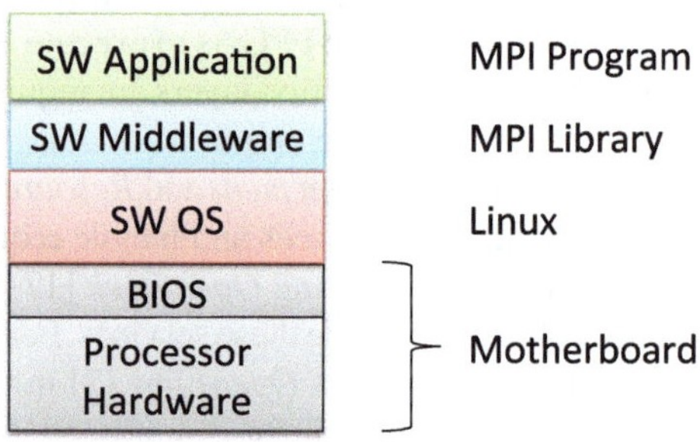

Fig. 2.1 Abstraction layers between software applications and the hardware (from [9])

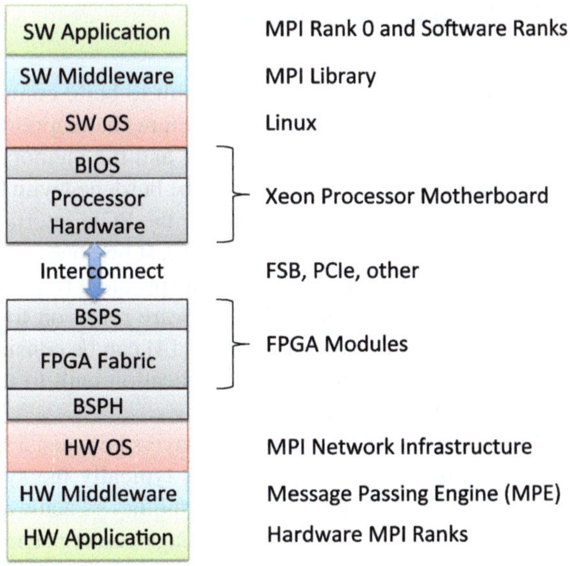

Fig. 2.2 Adding abstraction layers between hardware applications and the physical FPGA hardware to build a heterogeneous platform (from [9])

operating system, which is essentially a layer that provides some *services*, such as memory access and networking. In our MPI system [21], the on-chip networking and bridges to communicate off-chip would be considered to be in the HW OS layer. The *HW Middleware* layer provides a programming model or application interface to the hardware application. The Message Passing Engine [21] provides a hardware interface to message-passing functions used by our hardware MPI functions, so it would be considered a *HW Middleware* layer.

The abstractions shown in Fig. 2.2 make it possible to build heterogeneous applications that can be portable across many platforms as long as the requisite abstraction layers are provided. When moving to a data centre or cloud environment, there are even more layers of abstraction required. Figure 2.3 shows the additional abstraction layers that are needed to build a cloud environment. The software and hardware abstractions just discussed remain because they are needed to support the software and hardware applications on the software and hardware platforms, respectively. However, a cloud environment entails *scaling* to thousands of computing platforms, which in our work can be traditional processors, or servers, running software and platforms that comprise FPGAs to execute applications that are implemented in hardware. The cloud abstraction layers are required to support and manage the large-scale infrastructure that is the cloud.

The *Resource Management* and *Resource Allocation* layers are used to keep track of the computing resources and handle requests for resources from the users. In our environment, we are using *OpenStack* [17], which is an open-source cloud management platform to provide these services. For the networking layer, we use components of *OpenStack* as well as *OpenFlow* [2] in the SAVI Network testbed [3]. The interconnect in a cloud platform is a network. In Fig. 2.3, the *Interconnect* is shown to

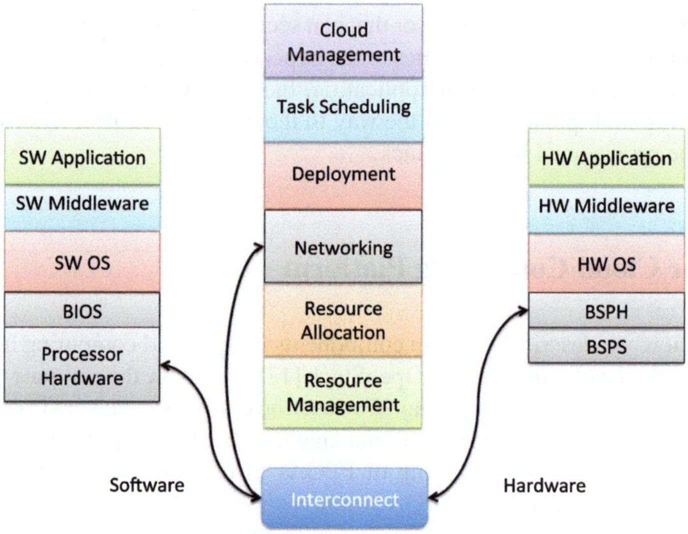

Fig. 2.3 Abstractions in a cloud environment

connect to the *Processor Hardware* where the *Network Interface Card* (NIC) resides and to the *BSPH* on the hardware platform, which contains the network interface. These interfaces are configured from the *Networking* layer in the cloud infrastructure. The *Deployment* layer is responsible for creating the virtual cluster configuration required for an application and loading the application into the computing platforms, which would mean loading bitstreams in the case that FPGAs are used. Components of *OpenStack* can do this for software applications, but not for hardware applications that are to run in FPGAs. Our approaches for deployment in two use cases are described in Sects. 2.4 and 2.5.

In a cloud environment, we typically talk about allocating and using *Virtual Machines* (VMs) rather than physical machines. *Containers* are also popular, but at the level of this discussion, we will treat them as a variation on the VM abstraction and focus the discussion on VMs. The abstraction of using virtual machines makes it possible to share physical machines by enabling them to support multiple virtual machines. To support this virtualization, we modify the software abstractions by inserting a *hypervisor* layer below the OS level. The hypervisor interfaces to the hardware and presents a virtual hardware interface to the guest operating systems that run on top of it thereby enabling the sharing of the hardware by the guest operating systems. The parallel on the hardware side of the abstractions also occurs at the OS, i.e. HW OS, layer. Given the limited types of functionality currently provided at the HW OS layer, we do not make the distinction between an OS and a hypervisor at this time. Instead, we will use the term *hypervisor* to represent the HW OS layer shown in Figs. 2.1 through 2.3. The hypervisor is equivalent to the *shell* in Microsoft terminology [19]. In Sect. 2.3.4, we describe a basic hypervisor and Sect. 2.6 introduces our current work towards a hypervisor that can securely support multiple tenants, i.e. multiple applications. The capabilities of the hypervisor have an impact on what kind of support can be provided to an application, the security and privacy capabilities for supporting multiple tenants, all the way to the higher level abstractions, such as what can be provisioned by OpenStack.

2.3 Our Cloud Computing Platform

In this section, we describe the main components in the cloud computing infrastructure that we are developing. We use OpenStack [17] to manage the compute resources and OpenFlow [2] for the networking. We describe our physical testbed and then our basic FPGA hypervisor that we use to manage user hardware applications and how this interfaces with our provisioning of FPGAs in the data centre. Then, we will describe how we provision the FPGA resources to allow other network-connected devices within the cloud to communicate with the FPGAs. This infrastructure is described in detail in [23]. We close this section by describing a software/hardware design flow for the cloud that can be used in our platform.

2.3.1 OpenStack

OpenStack [17] is an open-source cloud management platform that services user requests for virtual machines and maps the virtual resources onto physical resources in the data centre. A user request includes specifications of the required machine (or cluster) such as memory, hard-disk space and even PCIe devices (this is how we provision FPGAs in our environment), along with a software image that is to be loaded on the virtual machine. The physical details of the resources of the underlying physical server are abstracted away from the user. The physical resources in the data centre are all tracked by OpenStack. Such resources include computing resources, such as conventional CPU resources and FPGAs (through our extensions), and networking resources. For CPU resources, OpenStack communicates with a manager on each physical server referred to as an *agent*. The agent relays information pertaining to the resource utilization on the physical server, and OpenStack then uses this information to determine if the physical server can service the user's request.

2.3.2 Software-Defined Networking and OpenFlow

Our infrastructure includes both CPUs and FPGAs that each has their own direct connections to the network. To configure flexible network connections between CPUs and FPGAs, we use Software-Defined Networking (SDN). SDN is a concept that enables programmatic control of entire networks via an underlying software abstraction [16]. This is achieved by the separation of the network control plane from the data plane. SDN opens the door for users to test custom network protocols and routing algorithms, and furthermore, it allows the creation, deletion and configuration of network connections to be dynamic. The current de facto standard protocol for enabling SDN is OpenFlow [2].

SDN provides an abstraction layer for the user through an SDN controller. The SDN controller, often referred to as the *network operating system*, abstracts away network details from the user programs. These details include the physical locations of the switches and the configuration of the physical network. A user network program would be a program in the control path responsible for configuring a data path usually based on some input such as network usage. An example can be a load balancer, where abstractions are provided to view the network usage on the switches and the user program can redirect traffic accordingly. Switches in the data plane can simply handle header matching and basic packet modifications, however, parsing the payload or complex pattern matching requires a user program in the control path.

2.3.3 The SAVI Testbed

The Smart Applications on Virtualized Infrastructure (SAVI) testbed is a Canada-wide, multi-tier, heterogeneous testbed [3]. The testbed contains various heterogeneous resources such as FPGAs, GPUs, network processors, IoT sensors and conventional CPUs. The virtualization of these resources is one of the goals of SAVI, where our work investigates the FPGA platforms. Some resources, such as GPUs and network processors, are given to the user either by providing the entire machine without virtualization or with the use of PCIe passthrough, which is a mechanism for giving a virtual machine access to a physical PCIe device.

The multi-tier property refers to the network architecture of SAVI. SAVI can be seen as multiple cloud networks. The core network consists of a large number of CPUs that provide the backbone of the data centre. This core network is then connected to several edges dispersed around Canada. Each of these edges is a miniature cloud network that also contains the heterogeneous devices. Many of these heterogeneous devices are connected directly to the network through high-performance 10 Gb switches. These devices are treated the same as any CPU would be treated as many of these devices are assigned network ports with valid MAC and IP addresses. These devices are addressable by any other node (CPU or other device) on the network, once they are registered to the network. This allows, for example, an IoT sensor in Victoria to send data to an FPGA cluster in Toronto and then have the processed results be accessible by a CPU cluster in Calgary. Furthermore, the multi-tier architecture enables much of the processing to be done on the edge network with the heterogeneous devices before being sent to the large CORE where more compute resources are available.

From the perspective of our work, SAVI is a working cloud computing platform that uses OpenStack and OpenFlow. We leverage the existing SAVI testbed to develop our FPGA-based cloud infrastructure.

2.3.4 A Basic FPGA Hypervisor

In our initial cloud deployment, we have implemented a simple FPGA hypervisor that provides a basic I/O abstraction for a user application. The abstracted I/O includes the network, PCIe and off-chip DRAM. The basic FPGA hypervisor is shown in Fig. 2.4. Along with our hypervisor, we provide a driver called the Hardware Abstraction Layer (HAL) that can communicate to the FPGA via PCIe. The HAL provides an API for the user to communicate with various components in the FPGA through a memory-mapped interface. The communication includes configuring a processor in the hypervisor, sending control data to an application and transferring data to and from off-chip memory. With these connections, users can use the hypervisor to create a data path for their application using the network and configure control signals via the HAL.

Within the FPGA, the components controlled by the HAL are accessed using the AXI protocol. This is the standard protocol for memory-mapped interfaces used by Xilinx memory-mapped IP cores as well as ARM cores [7]. For the data path of our application, we use an AXI stream interface that is connected to the 10 Gb/s

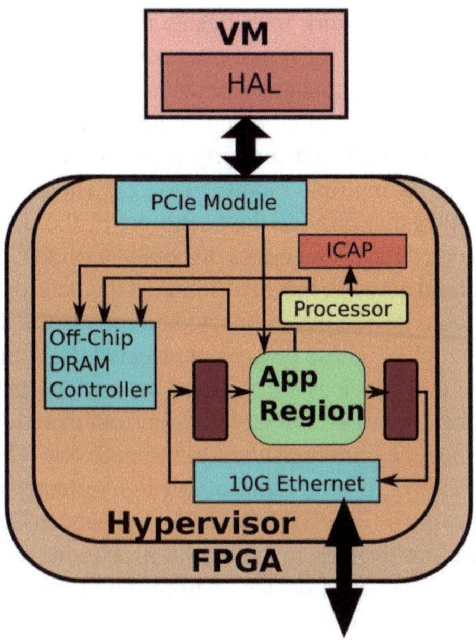

Fig. 2.4 System diagram of our basic FPGA hypervisor

networking interface on the FPGA. This interface streams packets as 8-byte words at a clock rate of 156.25 MHz.

The hardware application is programmed into the application region using partial reconfiguration. This ensures that only the application region is reprogrammed while keeping the rest of the hypervisor online and configured. The partial bitstream is transferred from the CPU to the FPGA using PCIe.

The programming and management of the application region are done with the Processor shown in Fig. 2.4. The Processor manages the decoupling gates (denoted by the boxes in front and behind the Application Region) and the ICAP within the FPGA. The ICAP is an IP within the FPGA Hypervisor that programs the FPGA with a partial bitstream. To ensure safe programming, the decoupling gates disconnect the Application Region from the Ethernet module during partial reconfiguration. Gating ensures that there are no transactions in flight in the data path (AXI stream from the network) or the control path (AXI from the PCIe) while we reprogram the Application Region. Once reprogrammed, the Processor un-gates the Application Region, thus reconnecting it to the data and control paths. The gating, programming and un-gating of the Application Region by the Processor are initiated through an API call from the CPU through PCIe and the HAL. The programming of the Application Region can be extended to be done over the network if we make a path to communicate with the Processor through the network.

2.3.5 OpenStack and OpenFlow Deployment of an FPGA Resource

Deploying an FPGA in the data centre requires establishing a means to provision the FPGA and then to connect the FPGA in the network. We use OpenStack to do provisioning and OpenFlow to make the network connections.

2.3.5.1 OpenStack FPGA Provisioning

The first step is to be able to provide an FPGA within a virtual machine. PCIe passthrough gives a virtual machine full access to a subset of PCIe devices within a physical compute node. This can allow us to have multiple virtual machines on top of the physical machine able to access different PCIe devices. The agent on the physical server registers a PCIe device by the unique PCIe vendor and device ID. If the PCIe device is provisioned to a virtual machine, the agent marks the PCIe device as in use and will not provision this device to any other virtual machine, giving the virtual machine full access to the PCIe device. This also allows us to control the CPU specifications of the virtual machine coupled to the PCIe device. In some use cases, we observe that the CPU is merely used to send control signals for configuration and programming of the FPGA whereas the data path uses the

10 Gb/s network connections. Here, a small CPU for the virtual machine is sufficient. Sections 2.4 and 2.5 will give examples of this scenario. In the case where a full FPGA development environment must be run on the virtual machine, we might want to configure a larger CPU for the virtual machine.

2.3.5.2 SAVI FPGA Network Registration

To provision the network port on the FPGA, we use OpenStack to reserve a virtual network port using the Neutron project in OpenStack [18]. Neutron maintains a database of all network ports provisioned within the data centre. A network port by our definition includes a MAC address and an IP address. We then use the SAVI SDN controller to register the physical port on the switch connected to the FPGA with the virtual network port provisioned with Neutron. This includes storing the mapping of the virtual to physical port in a database accessible by our SDN Controller. When a packet matching the physical port and the packet header defined by the virtual port is observed by the switch, it notifies the controller and then the switches are configured with the flows to forward all subsequent packets matching the packet header. In our environment, we have a single FPGA per physical server, and by determining which physical server is hosting the virtual machine with the PCIe-connected FPGA, we can infer the physical port on the switch that connects to the FPGA using our internal database. Once this port is registered, it is now accessible on the network by any other device in the data centre, including other virtual CPUs and FPGAs.

2.3.6 Software/Hardware Design Flow for the Cloud

We deployed our FPGA cloud service in May 2015. Since then it has been used by students within the University of Toronto as part of their own FPGA development environment.

Our infrastructure in the cloud also allows designers of large multi-FPGA network designs to approach their design with an incremental design flow. An example of this design flow is as follows:

1 Implement all parts of the design in software. Each function is an OpenStack image that contains a software application listening to the network port, performing a function and outputting to the port.
2 Implement and test each individual function as an FPGA-offloaded design, i.e. the inputs and the outputs of the FPGA function are connected to the host CPU to allow for easier debugging.
3 Swap the software-based function with the tested FPGA-based kernel.
4 If the heterogeneous cluster remains functionally correct, then repeat Steps 2 and 3 for the next function the chain. Repeat until the whole cluster is implemented using FPGAs.

This design flow allows for easy sharing of the physical FPGA. Cloud managers can track usage of the physical FPGAs by using monitoring functions provided by OpenStack.

2.4 A Middleware Platform for Building Heterogeneous FPGA Network Clusters

In this section, we describe a framework that allows users to map large logical clusters of IP blocks onto a physical multi-FPGA cluster [23]. Our middleware platform automatically divides a logical cluster into multiple FPGAs based on a mapping file, provides the logic for the network communication and issues the OpenStack and OpenFlow commands to provision and connect the clusters.

2.4.1 Logical-to-Physical Mapping of Kernels

We define a streaming digital circuit as a *kernel* within our infrastructure. The kernels in this system are streaming kernels using the AXI stream protocol for input and output.

All kernel inputs to the system are addressed by a specific destination entry in a packet. Logically speaking, unless otherwise stated, any kernel output can connect to any kernel input. This can be seen as all kernels being connected to a large logical switch. These kernels may be mapped to the same FPGA or to different FPGAs. Furthermore, these kernels can be replicated with directives in the input scripts and they can be scheduled in different ways with the use of schedulers.

Along with the logical cluster description of the circuit, the user provides a mapping file. The mapping file specifies the number of physical FPGAs the user requires for implementing their logical cluster and the assignment of kernels to the FPGAs. Given this mapping, the logical switch is mapped into several physical switches across multiple FPGAs. Furthermore, we create the logic to communicate between FPGAs by appropriately encapsulating packets with network headers (MAC address). Lastly, we program the network switches to ensure that all FPGAs can address every other FPGA with the appropriate MAC address. Figure 2.5 illustrates an example of a logical cluster being transformed into a physical cluster by our hardware middleware.

2.4.2 FPGA Application Region

The application region is shown in more detail in Fig. 2.6. It includes Input and Output helper modules for each user kernel to interface directly with the network

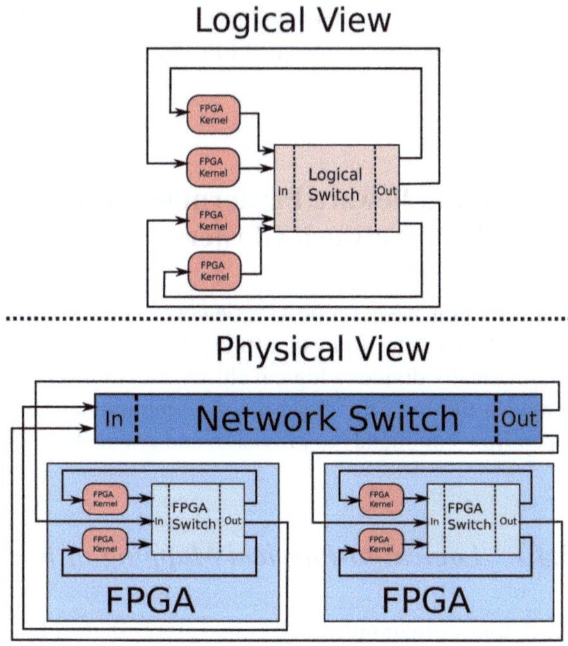

Fig. 2.5 This is an example of a logical cluster being transformed into a physical cluster. The user provides the logical cluster and mapping file (not shown), which is then transformed into a physical cluster. The FPGA switches, logic to communicate over the network (not shown) and the programming of the network switch is automatically generated

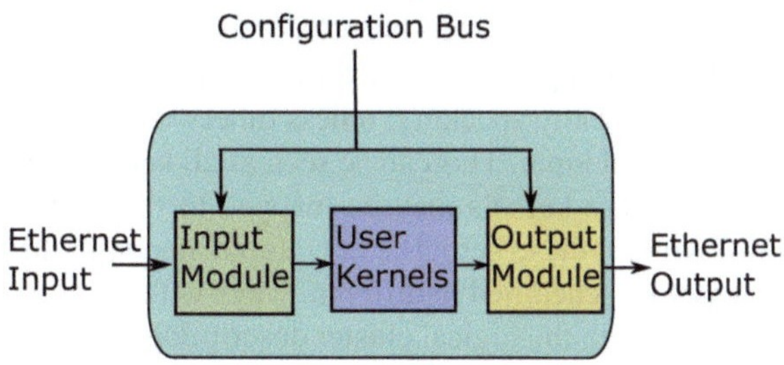

Fig. 2.6 Details of the application region. The input and output modules are both configured by the configuration bus (from [23])

through the Ethernet interface. The helper modules are responsible for filtering packets, formatting packets and arbitrating for the network port. The configuration bus is used to configure the input and the output modules. These signals are driven by the PCIe module in the Hypervisor shown in Fig. 2.4, which receives signals from the PCIe-connected virtual CPU.

The Input and Output modules are automatically configured by our middleware platform. This ensures that users need only consider the communication of kernels at a logical level and are not concerned with the inter-FPGA communication of these kernels if they are on different FPGAs. This also ensures that users do not have direct access to the Ethernet headers, which can be a security requirement within some data centres.

The Input Module is responsible for receiving packets from the network, filtering packets not destined for this FPGA (using the header) and then directing the packet to the destination kernel on the FPGA using a destination address. Each packet has two layers of addressing, one is the MAC address of the FPGA and the second is the kernel address within the cluster.

The Output Module is configured with a packet encapsulator to ensure the packet is received by the next hop. Our middleware platform is aware of the inter-FPGA connections and configures the Output Modules with the appropriate headers (FPGA MAC address and kernel address) to ensure that network packets are received by the appropriate FPGA and user hardware kernel.

2.4.3 The Creation of a Heterogeneous Network Cluster

We summarize the use of our system by describing the complete flow for building a heterogeneous network cluster of FPGAs. This flow can be seen as a middleware layer where the software component comprises scripts to parse input files, make calls to OpenStack and OpenFlow, and configure the hardware. The hardware component of this middleware layer includes all the additional hardware that is added to the application kernel to support the communication between kernels. In the first step, the user submits a logical cluster description and FPGA mapping file to a global FPGA parser. Our middleware then generates the appropriate OpenStack calls to provision the FPGAs and OpenFlow calls to connect the network-connected FPGAs. OpenStack calls are generated to create virtual machines. Most are lightweight CPU virtual machines connected to FPGAs plus one large virtual machine dedicated to synthesize bitstreams. Subsequent OpenStack calls are generated to create network ports, each with valid MAC and IP addresses. These ports are registered with the SAVI switch, and now all packets sent to these addresses will be forwarded to the correct switch port. After all the OpenStack calls are generated, the individual FPGAs are synthesized on the large virtual machine dedicated to synthesizing bitstreams. Once the bitstreams are synthesized, they are forwarded to the individual FPGAs to be programmed into the FPGA. Once programmed, the input and output modules are configured by the FPGA software driver running on each lightweight CPU attached to the FPGAs via PCIe. In summary, after the user submits the initial cluster description files, the rest of the calls are automatically generated by our middleware platform. The user receives a connected FPGA network cluster for their logical cluster and mapping file. This virtual abstraction hides all the automatically generated OpenStack calls, registration of network ports, bitstream generation and programming of the FPGAs.

2.5 A Middleware Platform for Integrating FPGAs into Virtualized Network Function Service Chains

This section describes a middleware platform that models FPGAs as Virtualized Network Functions (VNF) [22]. Section 2.4 describes a middleware platform where any FPGA kernel can communicate to any other kernel through a large logical switch potentially spanning many FPGAs. In service chaining, all the traffic through a function (implemented as software or a hardware kernel) is sent downstream through a specific path of functions. These can be seen as *middle boxes* for networking applications, each with a network source and sink. Each VNF can intercept the traffic, process the traffic and send the processed traffic to the network sink. This can be seen as a circuit switched network where the paths are stationary (unless reconfigured by our Service Chain Scheduler discussed in Sect. 2.5.3).

2.5.1 *Network Function Virtualization*

Network Function Virtualization (NFV) [11] is a concept for virtualizing network functions that have traditionally been provided by proprietary vendors as closed-box appliances. A network function is defined as a device that provides a well-defined network service, ranging from simple services, such as firewalls, content caches and load balancers, to more sophisticated ones such as intrusion detection and prevention. With recent gains in CPU performance, NFV aims to virtualize these services and create Virtualized Network Functions (VNFs) in software, running on commodity devices.

2.5.2 *Service Chaining of VNFs*

The activities surrounding SDN complement the recent work on NFVs. Both concepts aim to enable more flexible provisioning and management of infrastructure resources. When NFV is deployed in conjunction with SDN, it gives network operators the ability to create complex network services on demand by steering traffic through multiple VNFs realized using programmable resources. This practice is often called *service function chaining* (SFC), or simply *service chaining*. In addition, since NFV allows VNFs to be created or migrated closer to where they are needed, SDN can be leveraged to automatically re-route traffic to wherever the VNFs are placed.

The ability to share VNFs between different traffic flows can even be realized. Consider the case of email traffic that is steered through a service chain comprised

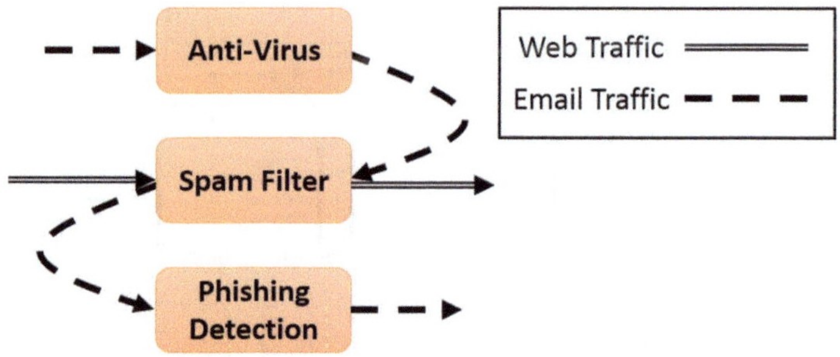

Fig. 2.7 An example of a service chain consisting of anti-virus, spam filter and phishing detection VNFs. Email traffic goes through a chain of all three VNFs, while web traffic only goes through one, which is shared with the email traffic (from [22])

of anti-virus, spam filtering and phishing detection VNFs. Similarly, web traffic can be steered through the same spam filtering VNF to block ads. An example is shown in Fig. 2.7, where an operator forms a chain of VNFs.

2.5.3 FPGA VNF Service Chain Scheduler

We use the service chain model to create heterogeneous chains of FPGAs and CPUs in our data centre. We present our service chain scheduler that we use to create heterogeneous network flows within our data centre. The service chain scheduler's main role can be described in two stages: the allocation stage and the networking stage. The allocation stage is responsible for provisioning the appropriate resources. This involves generating the OpenStack provisioning commands to acquire the FPGAs and the network ports, create the FPGA bitstreams and program the FPGA. The networking stage issues the OpenFlow commands to program all the switches to create the path specified by the user in their requested service chain. Figure 2.8 shows the high-level view of the service chain scheduler, where the two stages of its operation are illustrated.

2.5.4 Overview of Our VNF Middleware Platform

This is another example of a middleware platform that builds on top of our heterogeneous infrastructure. The platform is similar to the platform presented in Sect. 2.4 that automated the provisioning of resources in the cloud and created a cluster described by the user. The main difference is that this VNF middleware platform creates a fixed path that we can modify with our service chain scheduler. We can also leverage the incremental design flow that was described in Sect. 2.3.6, where we can create our

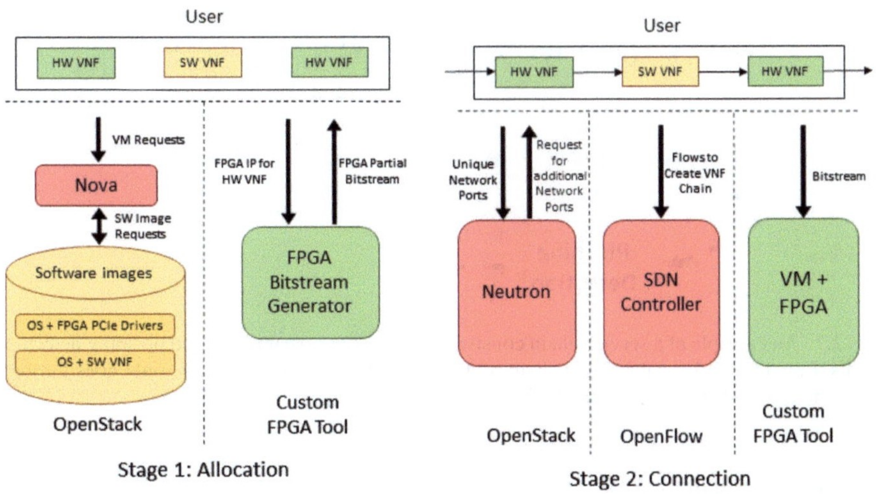

(a) The allocation stage of the Service Chain Scheduler.

(b) The connection stage of the Service Chain Scheduler.

Fig. 2.8 The service chain scheduler is divided into two stages, an allocation stage and a connection stage (from [22])

entire chain using CPU VNFs first and incrementally swap CPU VNFs for FPGA VNFs. This also allows service providers to consider different tiered VNF services. For example, they could implement a software implementation of their VNF for a low-cost service and an FPGA implementation of their VNF for a premium service.

2.6 A Multi-tenant Hypervisor

The hypervisor described in Sect. 2.3.4 enables a basic FPGA deployment where the FPGAs can be considered as peers to traditional compute nodes. The primary requirement is for an application thread to be able to interact directly with other threads over the network without requiring a host processor to broker the interaction. This is acceptable when we have a platform where physical computing devices interact directly with other physical computing devices. However, in a cloud environment, sharing of the physical computing devices by multiple users, or tenants, is common with processors and our goal is to support the same capability with FPGAs. The required capabilities must be provided by the hypervisor. This section describes ongoing work that is building towards an FPGA hypervisor that can support multiple tenants and preserve the abstractions we desire to make application development easier.

The sharing of resources between mutually distrusting hardware applications brings with it two key requirements: data isolation, the separation and protection of an application's data from other applications; and performance isolation, the decoupling of the performance of one application from the activities of another application. We manage these requirements by providing virtualized access to memory and networking resources.

In Sect. 2.2, we discussed the need for abstractions to make application development easier and portable. While the networking and memory access of the hypervisor presented in Sect. 2.3.4 constitutes a very basic and functional hardware OS with simple abstractions, we wish to implement a hypervisor with a richer set of abstractions, further easing the development of hardware applications while enabling the sharing of a physical FPGA by multiple tenants. In the remainder of this section, we discuss the design of an FPGA hypervisor that aims to implement these more advanced abstractions to provide true virtualization.

2.6.1 A Hypervisor Architecture Supporting True Virtualization

To facilitate a multi-tenant FPGA platform, Partial Reconfiguration (PR) is used to create multiple application regions on the FPGA, i.e. the FPGA is shared spatially. The use of PR makes it possible to program an application region without affecting other application tenants. When using PR, it is important to decide what logic is included in the *static region*, the part of the FPGA that surrounds the PR regions. Should all potential higher level abstractions be supported by the static region?

To help answer this question, recall from Sect. 2.2 that the purpose of an HW OS is to provide a common abstraction across a range of systems. An effective FPGA hypervisor should be designed such that the abstraction can be implemented on a broad range of PR regions, FPGA boards and even eventually FPGA vendors. A consequence of this need is that the lowest level of portability for a hardware application will be the application source code, such as Verilog, or perhaps a circuit netlist, given that bitstream portability is unlikely to be feasible in the foreseeable future. Therefore, higher level abstractions and services can also be implemented as source code or netlist-based IP cores that are *linked* to the application before the place and route step, and do not need to be included in the static region.

For the design of our hypervisor, we abide by the rule that only those components of the system that need to be shared among the PR application regions are included in the static region. From a virtualization perspective, each shared resource (i.e. memory, network connection, etc.) must implement data and performance isolation that is outside the reach of the end user, so the shared resources must be implemented in the static region.

Microsoft coined the term *shell* [19] for the static region in their FPGAs, which has now become commonly used by others. In our hypervisor, we introduce the concept of a *hard shell* and a *soft shell*. The hard shell provides only minimal features to facilitate multi-tenancy, which must also be inaccessible, i.e. not changeable by the

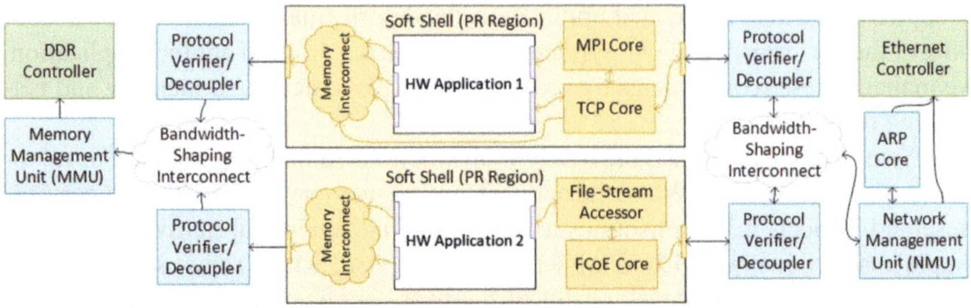

Fig. 2.9 System diagram of a multi-tenant FPGA hypervisor. Each application has a user-defined interface, with the soft shell implementing the necessary IP cores and glue logic to achieve the desired functionality. Performance isolation is ensured with bandwidth shaping interconnects and the memory/network management units

user. The soft shell is a generated wrapper for a user application that provides higher level abstractions and services specifically required by the user application.

Analogous to the soft shell is making system calls to an operating system for particular services. Since an FPGA has limited resources, it is not possible to include all possible services that might be required. Instead, we generate the soft shell to provide only the services required by an application, such as a TCP/IP stack or a particular memory interface. An augmented synthesis flow takes the user application and generates the set of IP cores that implement all the required higher level abstractions, plus the interconnect required to pair this newly created soft shell to the PR boundary interface of the hard shell. As shown in the hypervisor system diagram of Fig. 2.9, each application region can have its own unique soft shell.

2.6.2 *Memory Channel Virtualization*

One of the key resources used by hardware applications on FPGAs is off-chip memory. An important requirement we address is to ensure that the sharing of memory resources guarantees both performance and data isolation. To provide data isolation, a Memory Management Unit (MMU) is used to translate memory requests from the application regions to their allocated portion of memory. This MMU can either implement a segment-based remapping (some additive offset and max size) or a page-based remapping. If the page table is stored completely in on-chip SRAM (as in our current implementation), the tradeoff between the two types of remapping is only that of internal vs external memory fragmentation. Our hypervisor is parameterized to implement either form of remapping, though a study of which fragmentation problem is more significant for hardware applications is still needed and will be part of our future work.

An advanced feature we have implemented in our hypervisor is something we call *virtual channels*, where each logical memory interface instantiated in the application can be allocated a range of the memory region independently, with MMU translation occurring on a per logical interface basis, rather than having a single translation for the entire application region. From the user's point of view, each logical memory interface (declared in the user's source code) appears and acts as an independent memory channel. One example where this separation is necessary is shown in Application 1 of Fig. 2.9 where a TCP core instantiated in the soft shell needs access to off-chip memory. The TCP core is given a separately allocated part of memory that the user application cannot access and cannot corrupt.

To ensure performance isolation, we need to be able to regulate the memory bandwidth used by each application. This ensures that one application cannot spam the memory interface with requests that cause other applications to be denied access. Our hypervisor implements a bandwidth shaping interconnect to arbitrate access to shared resources and guarantee bandwidth allocation requirements. In addition to bandwidth allocation, illegal or spurious requests to any shared interconnect need to be blocked to prevent the interconnect from entering an error state or stalling operations. A protocol verifier and decoupler are included for this purpose, complete with timeout conditions.

2.6.3 *Networking Virtualization*

Our hypervisor also implements data and performance isolation for shared and virtualized networking resources. One possible way to implement performance isolation for network traffic is using the Enhanced Transmission Selection of IEEE 802.1Qaz [6] to implement traffic shaping in the Ethernet network while using the Priority Flow Control of IEEE 802.1Qbb [5] to assert back pressure on the network interfaces of each PR region. While this would offload the bandwidth shaping to the networking infrastructure of the data centre, it complicates the network setup and limits each PR region to a single IEEE P802.1p [4] class of service, which could be undesirable. Instead, a bandwidth shaping interconnect like that used for the memory interconnect is used to implement performance isolation. Note, the IEEE 802.1Qbb protocol can still be implemented to allow for higher level abstractions and functionality that require lossless Ethernet networking, such as Fibre Channel over Ethernet (FCoE).

Implementing data isolation for a networking interface is more complicated, since it depends on the level of abstraction at which the networking infrastructure is provided. If network isolation is ensured at the VLAN level, each application's network traffic is appended with the proper VLAN tag and traffic received from that VLAN is forwarded to that application. If isolation is ensured at the Ethernet layer (L2), the source and destination MACs are modified and traffic received for that MAC address is forwarded (i.e. each application has a unique MAC address). If isolation is ensured at the IP Layer (L3), the source and destination IP addresses are modified, ARP requests are sent to retrieve the proper MAC address for unknown destinations and traffic destined for that IP address is forwarded (i.e. each application has an assigned MAC and IP address). Finally, if isolation is ensured at the TCP/UDP Layer (L4), the source and destination ports are modified and traffic received on the assigned port is forwarded.

Rather than restrict the hardware application developers to a specific networking layer, our Network Management Unit (NMU) can implement any of the above-described functionalities. Like the virtual channels used for the memory virtualization, each logical network interface created in the application has an ID that is forwarded with any of its traffic to the NMU; the NMU uses this ID to determine the required level of network restriction. Note, the NMU does not perform any encapsulation, e.g. an application cannot send a raw TCP stream to the NMU and expect a proper Ethernet packet to be output. Instead, the encapsulating logic is generated in the soft shell region, and the NMU simply checks the packet fields to ensure sent packets are isolated to permitted sections of the network. This NMU design allows, for example, an implementation where one application is restricted to a specific VLAN but can send/receive for multiple physical addresses, while two other applications share a fixed MAC/IP address but send/receive data on different TCP/UDP ports. The memory and network performance isolation logic is shown in Fig. 2.9.

2.6.4 Management and Host Connectivity

A mechanism to manage the multi-tenant FPGA system, including setting up the parameters of the MMU and the NMU and programming the applications regions, is required. This is accomplished using a CPU running a lightweight software OS that allows remote connections (e.g. using *ssh*) or runs an agent that can communicate with OpenStack. Such a management layer can either be run on an on-chip CPU core (for SoC-type FPGA boards) or by a small control VM on a PCIe-connected host. This management layer would also need to manage the features implemented in the soft shell. Upon generation of the soft shell, an XML file containing the soft shell configuration is also generated. The XML file is passed to the management API along with the bitstream. Each application region includes a management interface that is used to configure the soft shell.

Though the FPGA hypervisor is mainly designed to be used in a stand-alone computing model, some applications may want host connectivity to implement a more traditional FPGA offload model. To implement this, a planned extension of the hypervisor includes Single Root I/O virtualization (SR-IOV) for optional host connectivity to the application regions. Under this model, the management interface is connected to a management VM through one PCIe Physical Function (PF), and a second PF implements many Virtual Functions (VF) that can be connected directly to the application regions and other software VMs on the same host. This would allow for the FPGA hypervisor to facilitate both stand-alone computing hardware applications and more traditional offload accelerator hardware applications.

2.7 Live Migration of FPGA Functions

On cloud platforms, live migration is the process of transferring the current state of a running Virtual Machine (VM) from one server to another. It is usually a seamless process in which the VM does not know or notice that it has been live migrated. Major cloud providers, like Google [12] and Rackspace [20], make use of live migration to protect their customers from hardware failures, to bring down a host for maintenance, and to perform load balancing. FPGAs in the Cloud are prone to similar issues of that of server hardware and will require a live migration like mechanism to move applications from one FPGA host to another with minimum disruption and downtime to the customer.

Live migration on FPGAs comes with a few challenges. One is having a mechanism to save the state of an FPGA in a quick and efficient manner. Unlike with a server VM, where the state is stored in off-chip memory and in a few CPU registers, the state of an FPGA application is stored in its off-chip memory, on-chip memory and thousands of flip-flops. The contents of off-chip memory can be saved and transferred using a technique used in server VMs called *precopying* [10]. One approach to save the internal state of an FPGA application is to use the readback functionality on FPGAs[1] to save the current state of all the flip-flops and internal memories into a bitstream. That bitstream is then transferred to the target FPGA for programming. This approach allows for seamless migration without notifying the customer; however, it is slow and requires the application to be paused, which can potentially violate the Service-Level Agreement (SLA). This approach also requires the handling of very low-level challenges, such as how to start and stop the parts of a design that work in different clock domains.

The current approach we are taking is to let the application save its own state. This approach is like cooperative multitasking for software tasks in an operating system. An application should have the most insight into what it needs to save to resume its state after migration. This is more efficient than the previous approach because there is no need to blindly save everything on the FPGA, only what is needed to restore the state. This approach could also allow for applications to be migrated between different FPGA devices. By letting the application save and restore its state to and from memory, the state of the application is no longer bound to the internal state of the FPGA, i.e. flip-flops and internal memories. By providing a standardized interface to and from memory, the application's source code can be used to produce several bitstreams to target multiple FPGA devices with each of these bitstreams being functionally equivalent. The state of the application can be restored by transferring the memory contents of that application from the source FPGA to the target FPGA. While a significant downside of this approach is that application developers will have to explicitly add state saving and restoring functionality, it is a more feasible approach at this time. Our first goal for FPGA migration is to achieve some level of functionality because many of the higher level issues will be the same independent of how the low-level state is saved.

[1] This feature is not available on all FPGAs.

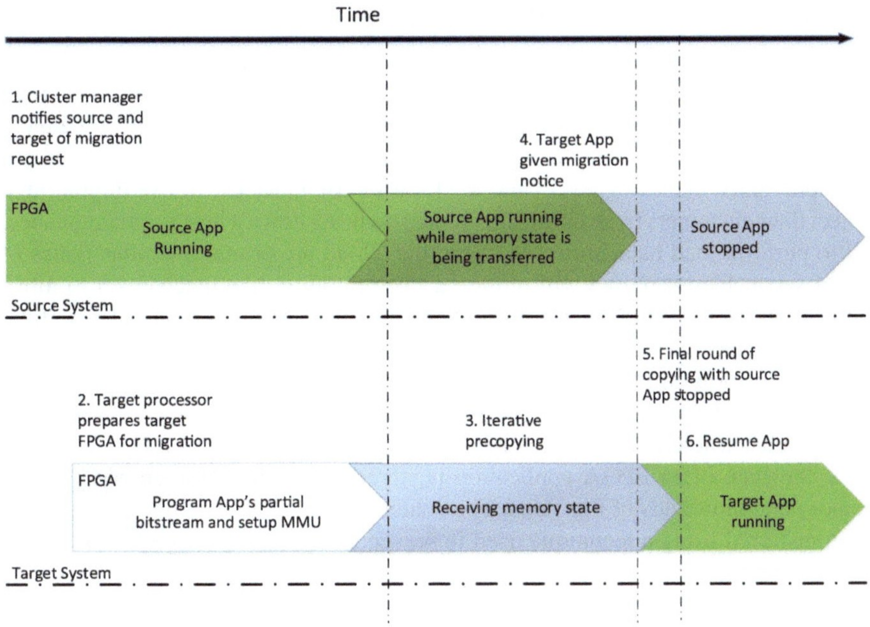

Fig. 2.10 Overview of the steps required to migrate an application between two FPGAs

An overview of the steps to migrate an application between two FPGAs is shown in Fig. 2.10. Applications are placed and operate within a Partially Reconfigurable (PR) region on the FPGA. An FPGA may have multiple PR regions for multiple applications. The loading of applications is managed by an on-chip processor or over PCIe by a host CPU. The migration process begins when the cluster manager schedules the migration task with the source and target processor. The target processor notifies the migration controller on the FPGA to set up the MMU to allocate the amount of memory specified by the manager. It will also program the PR with the target application's bitstream and put it into idle mode. The next phase involves copying the memory contents of the application from the source FPGA to the target FPGA using precopying. This is performed through the network port on the FPGA and not through the host processor. Some time during this phase, the source migration controller notifies the application of the pending migration to give the application time to save its state to off-chip memory. The application is then stopped, and a final round of copying is performed. The target migration controller sends a signal to the application to tell it to resume from a previous state, at which point the application will restore its state and resume operation on the target device. If an application's network configuration, e.g. IP and MAC addresses, will remain the same after the migration process, the network fabric will be reconfigured to send packets to the target device and any remaining packets destined for the application arriving at the source device will be forwarded to the target device.

2.8 Final Thoughts

The deployment of FPGAs in the cloud has already begun, starting with Microsoft [8, 19], recently Amazon Web Services [1] and others that have received less publicity. Based on the available information, most of the deployments treat the FPGAs as accelerators attached to a virtual machine except for Microsoft, where the FPGAs are directly connected to the network [8]. Amazon and Microsoft differ in another way. Amazon is making FPGAs generally available as a service, whereas Microsoft currently only uses their FPGAs as part of their infrastructure, so the FPGAs are not directly programmable by a customer. In the Amazon model, since the FPGAs are just attached to virtual machines, then the changes required in the Amazon cloud infrastructure are not too significant. Most likely, they can leverage most of what they use for providing their GPU service because the GPUs are also PCIe-connected devices. With the network-connected FPGAs that Microsoft is using the challenge is much greater. The work we are doing at the University of Toronto is much more relevant to the Microsoft model.

One of the most critical issues in the cloud is security. An FPGA sitting in a network can do a lot of damage because it can process and generate traffic at line rates. There are currently no inherent security mechanisms built into an FPGA, compared with processors where there are many features to support security. In the slave accelerator configuration that Amazon uses, the virtual machine can enforce the security around the FPGA, such as filtering packets from and to the network, so it is much easier to manage security. The main additional feature required is to guarantee the security and safety of the bitstreams. In the network-connected FPGA model used by Microsoft and in our platform, the security problem is much more challenging. That is probably the primary reason why Microsoft has not made their FPGAs programmable by their customers. Our heterogeneous network cluster hardware middleware we describe in Sect. 2.4 addresses some security concerns by restricting the user from directly sending packets to the network. The multi-tenant hypervisor we describe in Sect. 2.6 has features that continue moving towards providing security features to make generally accessible FPGAs secure. Eventually, to be truly robust, some of the features need to be in the hard logic of the FPGA, just like they are in modern CPU processors.

What Amazon and Microsoft both need to provide is a common hypervisor infrastructure, like we are proposing, to make user applications portable across platforms. If such a hypervisor existed, then it would be possible to build middleware that can support applications and make FPGAs much easier to program as well as make it easier to port applications between platforms. In our work, we have shown that the hypervisor is really the foundation that supports all of the abstraction layers above it, all the way into the cloud infrastructure, so the hypervisor design must be well considered.

Microsoft has already demonstrated the benefits of having network-connected FPGAs because they can be provisioned just like the CPUs. This means that the FPGAs can also be used as a resource that can scale with user demand, which is exactly what is required in the cloud. We see this as validation for our approach of integrating FPGAs into the cloud as peers to the processors. We have shown the benefits of abstractions and middleware to help build large-scale, multi-FPGA applications. There is still much work to do to make FPGAs easier to use and fully accessible in the cloud.

References

1. (2018) Amazon EC2 F1 Instances. https://aws.amazon.com/ec2/instance-types/f1/. Accessed 15 June 2018
2. (2018) OpenFlow. https://www.opennetworking.org/sdn-resources/openflow. Accessed 15 June 2018
3. (2018) Smart Applications on Virtual Infrastructure. https://www.savinetwork.ca/. Accessed 15 June 2018
4. (2004) IEEE standard for local and metropolitan area networks: media access control (MAC) bridges–annex G user priorities and traffic classes. IEEE Std 8021D-2004 (Revision of IEEE Std 8021D-1998), pp 264–268. https://doi.org/10.1109/IEEESTD.2004.94569
5. (2011) IEEE standard for local and metropolitan area networks–media access control (MAC) bridges and virtual bridged local area networks–amendment 17: priority-based flow control. IEEE Std 8021Qbb-2011 (Amendment to IEEE Std 8021Q-2011 as amended by IEEE Std 8021Qbe-2011 and IEEE Std 8021Qbc-2011), pp 1–40. https://doi.org/10.1109/IEEESTD.2011.6032693
6. (2011) IEEE standard for local and metropolitan area networks–media access control (MAC) bridges and virtual bridged local area networks–amendment 18: enhanced transmission selection for bandwidth sharing between traffic classes. IEEE Std 8021Qaz-2011 (Amendment to IEEE Std 8021Q-2011 as amended by IEEE Std 8021Qbe-2011, IEEE Std 8021Qbc-2011, and IEEE Std 8021Qbb-2011), pp 1–110. https://doi.org/10.1109/IEEESTD.2011.6034507
7. ARM AMBA (2003) AXI protocol specification
8. Caulfield A, et al (2016) A cloud-scale acceleration architecture. In: Proceedings of the 49th annual IEEE/ACM international symposium on microarchitecture, IEEE Computer Society. https://www.microsoft.com/en-us/research/publication/configurable-cloud-acceleration/
9. Chow P (2016) An open ecosystem for software programmers to compute on FPGAs. In: Third international workshop on FPGAs for software programmers (FSP 2016), 11 pp
10. Clark C, Fraser K, Hand S, Hansen JG, Jul E, Limpach C, Pratt I, Warfield A (2005) Live migration of virtual machines. In: Proceedings of the 2nd conference on symposium on networked systems design and implementation-volume 2, USENIX Association, pp 273–286
11. ETSI (2018) Network functions virtualisation (NFV); Architectural framework. https://portal.etsi.org/nfv. Accessed 15 June 2018
12. Google (2018) Google compute engine uses live migration technology to service infrastructure without application downtime. https://cloudplatform.googleblog.com/2015/03/Google-Compute-Engine-uses-Live-Migration-technology-to-service-infrastructure-without-application-downtime.html. Accessed 15 June 2018
13. Gropp W, Lusk E, Doss N, Skjellum A (1996) A high-performance, portable implementation of the MPI message passing interface standard. Parallel Comput 22(6):789–828. https://doi.org/10.1016/0167-8191(96)00024-5
14. Linux (2018) The Linux Foundation. http://www.linuxfoundation.org/. Accessed 15 June 2018
15. Madill, C., Patel, A., Saldaña, M., Pomès, R., Chow, P: A heterogeneous architecture for biomolecular simulation. In: Gaillardon PE (ed) Reconfigurable logic: architecture, tools, and applications, CRC Press, chap 11, pp 323–350 (2015)
16. McKeown N (2009) Software-defined networking. INFOCOM Keynote Talk

17. OpenStack (2018) OpenStack. https://www.openstack.org/. Accessed 15 June 2018
18. OpenStack Inc (2018) Welcome to Neutron's developer documentation! https://docs.openstack.org/developer/neutron/. Accessed 15 June 2018
19. Putnam A, et al (2014) A reconfigurable fabric for accelerating large-scale datacenter services. In: 41st annual international symposium on computer architecture (ISCA)
20. Rackspace (2018) Live migration overview. https://support.rackspace.com/how-to/live-migration-overview/. Accessed 15 June 2018
21. Saldaña M, Patel A, Madill C, Nunes D, Wang D, Styles H, Putnam A, Wittig R, Chow P (2010) MPI as a programming model for high-performance reconfigurable computers. ACM Trans Reconf Technol Syst 3(4):22:1–22:29
22. Tarafdar N, Lin T, Eskandari N, Lion D, Leon-Garcia A, Chow P (2017) Heterogeneous virtualized network function framework for the data center. In: Field programmable logic and applications (FPL). IEEE
23. Tarafdar N, Lin T, Fukuda E, Bannazadeh H, Leon-Garcia A, Chow P (2017) Enabling flexible network FPGA clusters in a heterogeneous cloud data center. In: Proceedings of the 2017 ACM/SIGDA international symposium on field-programmable gate arrays. ACM, pp 237–246

Chapter 3
dReDBox: A Disaggregated Architectural Perspective for Data Centers

Nikolaos Alachiotis, Andreas Andronikakis, Orion Papadakis, Dimitris Theodoropoulos, Dionisios Pnevmatikatos, Dimitris Syrivelis, Andrea Reale, Kostas Katrinis, George Zervas, Vaibhawa Mishra, Hui Yuan, Ilias Syrigos, Ioannis Igoumenos, Thanasis Korakis, Marti Torrents and Ferad Zyulkyarov

3.1 Introduction

Data centers have been traditionally defined by the physical infrastructure, imposing a fixed ratio of resources throughout the system. A widely adopted design paradigm assumes the mainboard tray, along with its hardware components, as the basic building block, requiring the system software, the middleware, and the application stack to treat it as a monolithic component of the physical system. The proportionality of resources per mainboard tray, however, which is set at design time, remains fixed throughout the lifetime of a data center, imposing various limitations to the overall data center architecture. First, the proportionality of resources at system level inevitably follows the fixed resource proportionality of the mainboard tray. With the mainboard tray as the basic, monolithic building block, system-level upgrades to facilitate increased resource requirements for additional memory, for instance, bring along parasitic capital and operational overheads that are caused by the rest of the hardware components on a tray, e.g., processors and peripherals. Second, the process of assigning/allocating resources to Virtual Machines (VMs) is significantly

N. Alachiotis · D. Theodoropoulos · D. Pnevmatikatos
Foundation for Research and Technology - Hellas, Heraklion, Greece

N. Alachiotis · A. Andronikakis · O. Papadakis · D. Theodoropoulos · D. Pnevmatikatos
Technical University of Crete, Chania, Greece

D. Syrivelis · A. Reale · K. Katrinis (✉)
IBM Research, Mulhuddart, Ireland
e-mail: katrinisk@ie.ibm.com

G. Zervas · V. Mishra · H. Yuan
University College London, London, UK

I. Syrigos · I. Igoumenos · T. Korakis
University of Thessaly, Volos, Greece

M. Torrents · F. Zyulkyarov
Barcelona Supercomputing Center, Barcelona, Spain

C. Kachris et al. (eds.), *Hardware Accelerators in Data Centers*,
https://doi.org/10.1007/978-3-319-92792-3_3

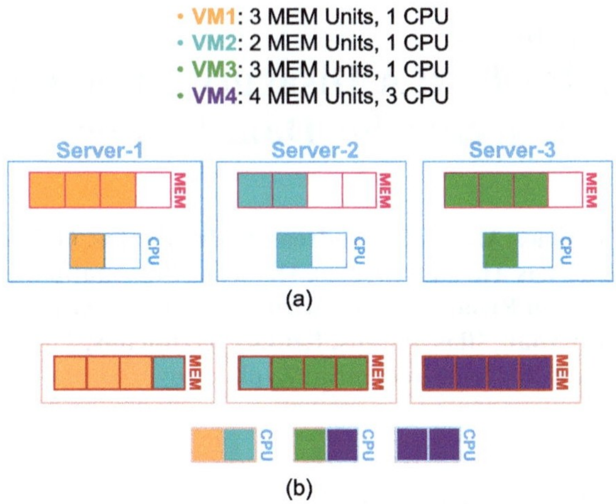

Fig. 3.1 A limitation of current data center infrastructures regarding resource utilization (**a**), and the respective resource allocation scheme of dReDBox (**b**)

restricted by the physical boundaries of the mainboard tray, with respect to resource quantities. Such lack of flexibility in allocating resources to VMs eventually brings the overall system to a state of stranded resource capacity, while yielding the data center insufficient to facilitate the requirements of additional VMs. Third, technology upgrades need to be carried out on a per-server basis, raising significantly the costs when applied at scale, e.g., in an Internet-scale data center that can potentially comprise thousands of servers (Fig. 3.1).

However, as current data center systems are composed by a networked collection of monolithic building blocks, shifting toward an architectural paradigm that overcomes the aforementioned fixed proportionality of resources by breaking the boundaries of the motherboard tray to achieve finer granularity in resource allocation to VMs entails various challenges and open problems to address. The challenges and requirements broadly relate to four categories, namely the hardware platform, the memory, the network, and the system software. A representative hardware-platform-level requirement, for instance, entails the need to establish intra- and inter-tray interconnection paths that are programmable, yet introduce minimal communication overhead. For this purpose, a low-latency network architecture that is scalable and achieves high bandwidth is needed. Remote memory allocation support and management is required, as well as efficient ways to maintain coherency and consistency, while minimizing remote-memory access latency. Dedicated mechanisms to support dynamic on-demand network connectivity and scalability, as well as orchestration software tools that define resource topologies, generate and manage VMs, and ensure reliability and correctness while exploring optimizations are prerequisites for the success of the overall approach. Fine-grained power management at the component level is required in order to minimize overall data center energy consumption.

To this end, the dReDBox (disaggregated Recursive Datacenter in a Box) project aims at overcoming the issue of fixed resource proportionality in next generation, low-power data centers by departing from the mainboard-as-a-unit paradigm and enabling disaggregation through the concept of function-block-as-a-unit. The following section (Sect. 3.2) provides an overview of existing data center architectures and related projects, and presents the dReDBox approach to data center architecture design. The remaining sections are organized as follows. Sections 3.3 and 3.4 present the system architecture and the software infrastructure, respectively. Section 3.5 describes a custom simulation environment for disaggregated memory and present a performance evaluation. Finally, Sect. 3.6 concludes this chapter.

3.2 Disaggregation and the dReDBox Perspective

The concept of data center disaggregation regards resources as independent homogeneous pools of functionalities across multiple nodes. The increasingly recognized benefits of this design paradigm have motivated various vendors to adopt the concept, and significant research efforts are currently conducted toward that direction, both industrial and academic ones.

From a storage perspective, disaggregation of data raises a question regarding where should data reside, at the geographical level, to achieve short retrieval times observed by the end users, data protection, disaster recovery, and resiliency, as well as to ensure that mission-critical criteria are met at all times. Various industrial vendors provide disaggregated solutions for that purpose, enabling flash capacity disaggregation across nodes, for instance, or the flexible deployment and management of independent resource pools. Klimovic et al. [13] examine disaggregation of PCIe-based flash memory as an attempt to overcome overprovisioning of resources that is caused by the existing inflexibility in deploying data center nodes. The authors report a 20% drop in application-perceived throughput due to facilitating remote flash memory accesses over commodity networks, achieving, however, highly scalable and cost-effective allocation of processing and flash memory resources.

A multitude of research efforts has focused on disaggregated memory with the aim to enable scaling of memory and processing resources at independent growth paces. Lim et al. [14, 15] present the "memory blade" as an architectural approach to introduce flexibility in memory capacity expansion for an ensemble of blade servers. The authors explore memory-swapped and block-access solutions for remote access, and address software- and system-level implications by developing a software-based disaggregated memory prototype based on the Xen hypervisor. They find that mechanisms which minimize the hypervisor overhead are preferred in order to achieve low-latency remote memory access.

Tu et al. [24] present Marlin, a PCIe-based rack area network that supports communication and resource sharing among disaggregated racks. Communication among nodes is achieved via a fundamental communication primitive that facilitates direct memory accesses to remote memory at the hardware level, with PCIe and Ethernet links used for inter-rack and intra-rack communication, respectively. Dragojević et al. [9] describe FaRM, a main memory distributed computing platform that, similarly to Marlin, relies on hardware-support for direct remote memory access in order to reduce latency and increase throughput.

Acceleration resources, e.g., FPGAs, are increasingly being explored to boost application performance in data center environments. Chen et al. [8] present a framework that allows FPGA integration into the cloud, along with a prototype system based on OpenStack, Linux-KVM, and Xilinx FPGAs. The proposed framework addresses matters of resource abstraction, resource sharing among threads, interfacing with the underlying hardware, and security of the host environment. Similarly, Fahmy et al. [10] describe a framework for accelerator integration in servers, supporting virtualized resource management and communication. Hardware reconfiguration and data transfers rely on PCIe, while software support that exposes a low-level API facilitates FPGA programming and management. Vipin and Fahmy [25] present DyRACT, an FPGA-based compute platform with support for partial reconfiguration at runtime using a static PCIe interface. The DyRACT implementation is targeting Virtex 6 and Virtex 7 FPGAs, while a video-processing application that employs multiple partial bitstreams is used as a case study for validation and evaluation purposes.

More recently, Microsoft presented the Configurable Cloud architecture [7], which introduces reconfigurable devices between network switches and servers. This facilitates the deployment of remote FPGA devices for acceleration purposes, via the concept of a global pool of acceleration resources that can be employed by remote servers as needed. This approach to disaggregation eliminates the, otherwise, fixed one-to-one ratio between FPGAs and servers, while the particular FPGA location, between the server and the network switches, enables the deployment of reconfigurable hardware for infrastructure enhancement purposes, e.g., for encryption and decryption. A prior work by Microsoft, the Catapult architecture [21], relied on a dedicated network for inter-FPGA communication, therefore, raising cabling costs and management requirements. Furthermore, efficient communication among FPGAs was restricted to a single rack, with software intervention required to establish inter-rack data transfers.

EU-funded research efforts, such as the Vineyard [12] and the ECOSCALE [17] projects, aim at improving performance and energy efficiency of compute platforms by deploying accelerators. The former addresses the problem targeting data center environments, via the deployment of heterogeneous systems that rely on data-flow engines and FPGA-based servers, while the latter adopts a holistic approach toward a heterogeneous hierarchical architecture that deploys accelerators, along with a hybrid programming environment based on MPI and OpenCL.

The dReDBox approach aims at providing a generic data center architecture for disaggregation of resources of arbitrary types, such as processors, memories, and FPGA-based accelerators. Basic blocks, dubbed bricks, construct homogeneous pools of resources, e.g., a compute pool comprising multiple compute bricks, with system software and orchestration tools implementing software-defined virtual

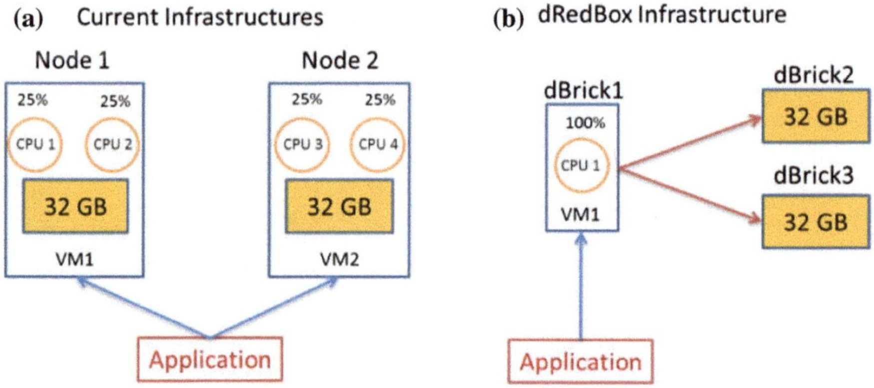

Fig. 3.2 Example of a memory-intensive application and the respective resource allocation schemes in a current infrastructure (**a**) and dReDBox (**b**)

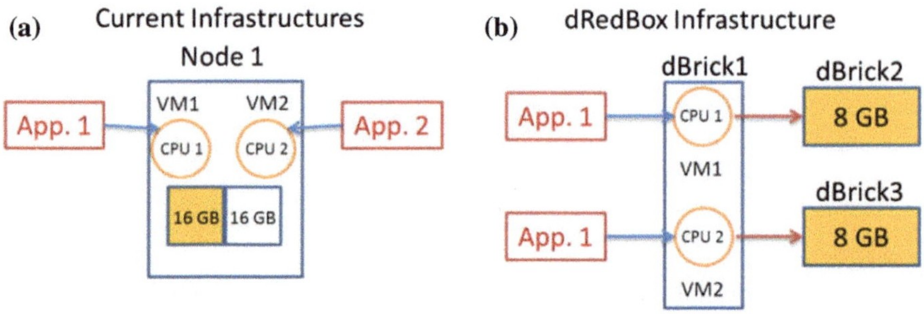

Fig. 3.3 Example of a compute-intensive application and the respective resource allocation schemes in a current infrastructure (**a**) and dReDBox (**b**)

machines which exhibit customized amounts of resources that better serve application needs. High-speed, low-latency optical and electrical networks will establish inter- and intra-tray communication among bricks, respectively. Figures 3.2 and 3.3 illustrate the expected resource allocation schemes, enabled by the dReDBox infrastructure, for serving the requirements of memory- and compute-intensive applications, respectively. The following section presents the proposed dReDBox hardware architecture.

3.3 System Architecture

This section presents the overall architecture of a dReDBox data center. Multiple dReDBox racks, interconnected via an appropriate data center network, form a dReDBox data center. This overall architecture is shown in Fig. 3.4. The dReDBox architecture comprises pluggable compute/memory/accelerator modules (termed bricks

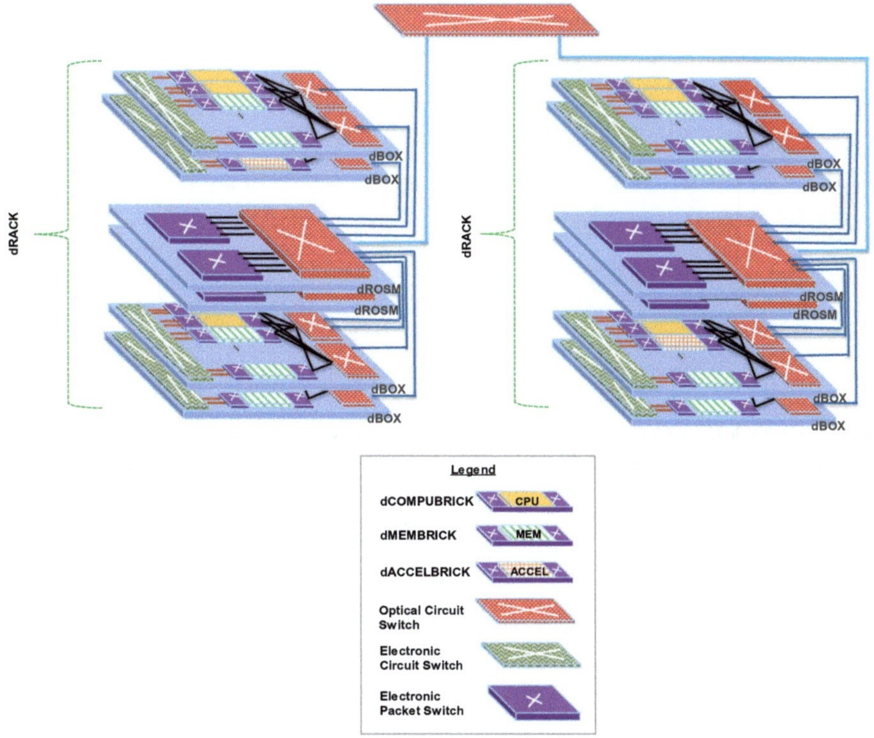

Fig. 3.4 Overview of a dReDBox rack architecture comprising several dBOXes interconnected with hybrid optical and electrical switching (dROSM)

in dReDBox terminology) as the minimum field-replaceable units. A single or sets of multiples of each brick type forms an IT resource pool of the respective type. A mainboard tray with compatible brick slots and on-board electrical crossbar switch, flash storage, and baseboard management components is used to support up to 16 bricks. A 2U carrier box (dBOX, visually corresponding from the outside to a conventional, rack-mountable data center server) in turn hosts the mainboard tray and the intra-tray optical switch modules.

3.3.1 The dBRICK Architecture

The dBRICK is the smallest realization unit in the dReDBox architecture. The term encompasses general-purpose processing (dCOMPUBRICK), random-access memory (dMEMBRICK), and application-specific accelerators (dACCELBRICK). As described above, dBRICKs will be connected to the rest of the system by means of a tray that, besides connectivity, will also provide the necessary power to each brick.

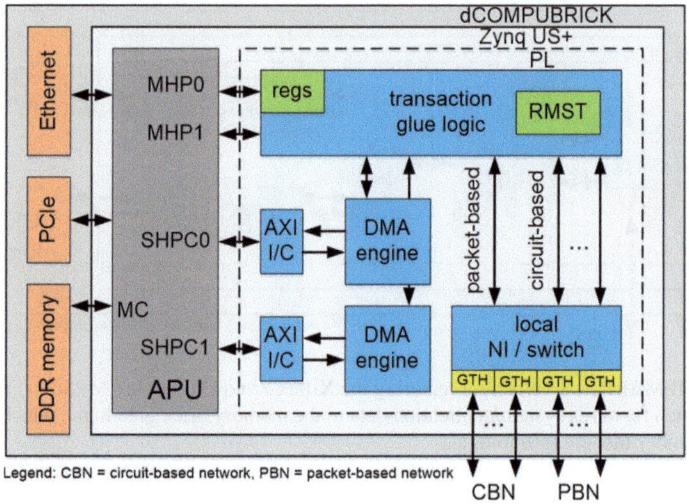

Fig. 3.5 Block diagram of a dCOMPUBRICK. The MPSoC integrates an Application Processing Unit (APU) for software execution. The on-chip programmable logic on the SoC is used to host transaction glue logic, housekeeping state, and communication logic, required for accessing disaggregated resources. The local DMA engines allow the system software to efficiently migrate pages from remote memory regions to local DDR memory

3.3.1.1 Compute Brick Architecture (dCOMPUBRICK)

The dReDBox compute brick (Fig. 3.5) is the main processing block in the system. It hosts local off-chip memory (DDR4) for low-latency and high-bandwidth instruction read and read/write data access, as well as Ethernet and PCIe ports for data and system communication and configuration. Also, each dCOMPUBRICK features QSPI-compatible flash storage (16–32 MB) and a micro-SD card socket (not shown in Fig. 3.5) to facilitate reset and reconfiguration of the brick in the case of disconnection, as well as for debugging purposes.

The compute brick can reach disaggregated resources, such as memory and accelerators, via dReDBox-specific glue intellectual property (termed Transaction Glue Logic) on the data path and communication endpoints implemented on the programmable logic of the dCOMPUBRICK MPSoC. System interconnection to disaggregated resources occurs via multiple ports leading to circuit-switched tray- and rack-level interconnects. As also shown in Fig. 3.5, we also experiment with packet-level system/data interconnection, using Network Interface (NI) and a brick-level packet switch (also implemented on programmable logic of the dCOMPUBRICK MPSoC), on top of the inherently circuit-based interconnect substrate. There is potential value in such an approach, specifically in terms of increasing the connectivity of a dCOMPUBRICK due to multi-hopping and thus creating an opportunity to increase the span of resource pools reachable from a single dCOMPUBRICK.

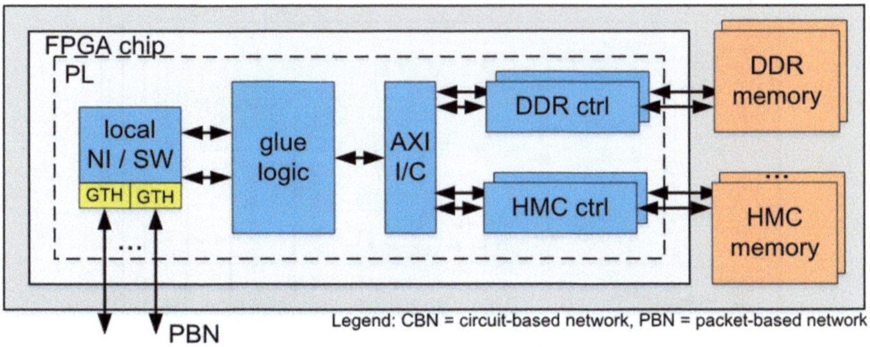

Fig. 3.6 dMEMBRICK architecture featuring the Xilinx Zynq Ultrascale+ MPSoC (EG version); the local switch forwards system/application data to the memory brick glue logic, which interfaces different memory module technologies

3.3.1.2 Memory Brick Architecture (dMEMBRICK)

Figure 3.6 illustrates the memory brick (dMEMBRICK) architecture, which is a key disaggregation feature of dReDBox. It will be used to provide a large and flexible pool of memory resources, which can be partitioned and (re)distributed among all processing nodes (and corresponding VMs) in the system. dMEMBRICKs can support multiple links. These links can be used to provide higher aggregate bandwidth, or can be partitioned by the orchestrator and assigned to different dCOMPUBRICKs, depending on the resource allocation policy used. This functionality can be used in two ways. First, the nodes can share the memory space of the dMEMBRICK, implementing essentially a shared memory block (albeit shared among a limited number of nodes). Second, the orchestrator can also partition the memory of the dMEMBRICK, creating private partitions for each client. This functionality allows for fine-grained memory allocation. It also requires translation and protection support in the glue logic (transaction glue logic block) of the dMEMBRICK.

The glue logic implements memory translation interfaces with the requesting dCOMPUBRICKs, both of which are coordinated by the orchestrator software. Besides network encapsulation, the memory translator, managed by orchestrator tools, controls the possible sharing of the memory space among multiple dCOMPUBRICKs, enabling support for both sharing among and protection between dCOMPUBRICKs. The control registers allow the local mapping of external requests to local addresses to allow more flexible mapping and allocation of memory.

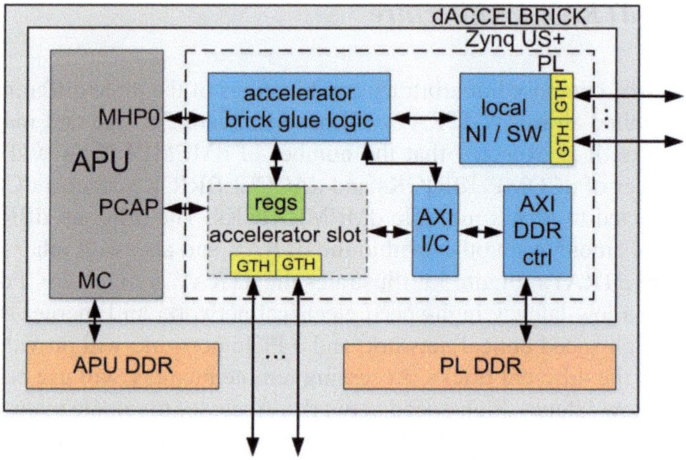

Fig. 3.7 The dACCELBRICK architecture for accommodating application-specific accelerators

3.3.1.3 Acceleration Brick Architecture (dACCELBRICK)

A dACCELBRICK hosts accelerator modules that can be used to boost application performance based on a near-data processing scheme [20]; instead of transmitting data to remote dCOMPUBRICKs, certain calculations can be performed by local accelerators, thus improving the performance while reducing network utilization.

Figure 3.7 depicts the dACCELBRICK architecture. The dACCELBRICK consists of the dynamic and the static infrastructure. The dynamic infrastructure consists of a predefined, reconfigurable slot in the PL that hosts hardware accelerators. As depicted in Fig. 3.7, the accelerator wrapper template integrates a set of registers that can be accessed by the glue logic to monitor and control (e.g., debug) the accelerator. Moreover, the wrapper provides a set of high-speed transceivers (e.g., GTHs) for direct communication between the accelerator and other external resources. Finally, an AXI-compatible port interfaces directly with an AXI DDR controller, allowing the hardware accelerator to utilize the local PL DDR memory during data processing. The static infrastructure hosts require modules for (a) supporting dynamic hardware reconfiguration, (b) interfacing with the hardware accelerator, and (c) establishing communication with remote dCOMPUBRICKs. To support hardware reconfiguration, in the current implementation, the local APU executes a thin middleware responsible for (a) receiving bitstreams from remote dCOMPUBRICKs (through the accelerator brick glue logic), (b) storing bitstreams in the APU DDR memory, and (c) reconfiguring the PL with the required hardware IP via the PCAP port. To monitor/control the hardware accelerator, the glue logic can read/write the wrapper registers. In addition, the glue logic interfaces with the local NI/switch for data transfers between the dACCELBRICK and remote dCOMPUBRICKs.

3.3.2 The dTRAY Architecture

dTRAYs may be composed of arbitrary combinations of the three different types of dBRICKs detailed above. A dTRAY will have standard 2U size and may contain up to 16 bricks. It is expected that the number of dMEMBRICKs will be larger than the number of dCOMPUBRICKs and dACCELBRICKs, since a dCOMPUBRICK is expected to access multiple dMEMBRICKs. The different dBRICKs are interconnected among each other within the dTRAY and also with other dBRICKs from different dTRAYs. Figure 3.8 illustrates the dTRAY architecture. Four different networks, a low-latency high-speed electrical network, an Ethernet network, a low-latency high-speed optical network, and a PCIe network, will provide connectivity between the different bricks. Accessing remote memory will use both optical and electrical low-latency, high-speed networks. Accesses to remote memory placed in a dMEMBRICK within a dTRAY will be implemented via an electrical circuit crossbar switch (dBESM in Fig. 3.4 is labeled as High-Speed Electrical Switch) and will connect directly to the GTH interface ports available on the programmable logic of the bricks. The dBESM switch will have 160 ports. This is the largest dBESM switch available on the market today supporting our speed requirements. The latency will be as low as 0.5 ns and the bandwidth per port will be 12 Gbps. This network will be used for intra-tray memory traffic between different bricks inside the tray. dBESM will not be used for inter-tray memory traffic due the limitations of the electrical communication in larger distances (latency). In addition, using electrical network for intra-tray communication instead of an optical network would not require signal conversion from electrical to optical and vice versa and thus it will be lower latency and lower power consumption. The optical network on the dTRAY, providing inter-tray connectivity, will be implemented with multiple optical switch modules (dBOSM in dReDBox terminology). Each dBOSM switch will have 24 optical ports. The latency of the dBOSM optical switch would be around 5 ns and the bandwidth would be in the range of 384 Gbps. dBRICKs will connect to the dBOSM via GTH interface ports available on the programmable logic of the SoC. The GTH bandwidth is 16 Gbps. A total of 24 GTH ports will be available in the SoC, with 8 of them used to connect the SoC to the dBOSM. On a fully populated tray hosting 16 bricks, a maximum of 256 optical ports may be used to fully interconnect the bricks of each tray. A Mid-Board Optics (MBO) device mounted on each dBRICK will be used to convert the electrical signals coming from the GTH ports and aggregate them into a single-fiber ribbon; the other end of the ribbon will be attached to a local dBOXs dBOSM optical switch. Each MBO supports up to eight ports. The dBRICKs will use 10 GTH ports to connect to dBOSM. The number of GTHs per SoC connecting to the dBOSM is limited by the size of the dBOSM. A 160-port dBOSM can support a maximum of 10 GTH per dBRICK, given a maximum of 16 dBRICKs on a tray. An Ethernet (ETH) network will be used for regular network communication and board management communication (BMC). The bandwidth will be 1 Gbps and it

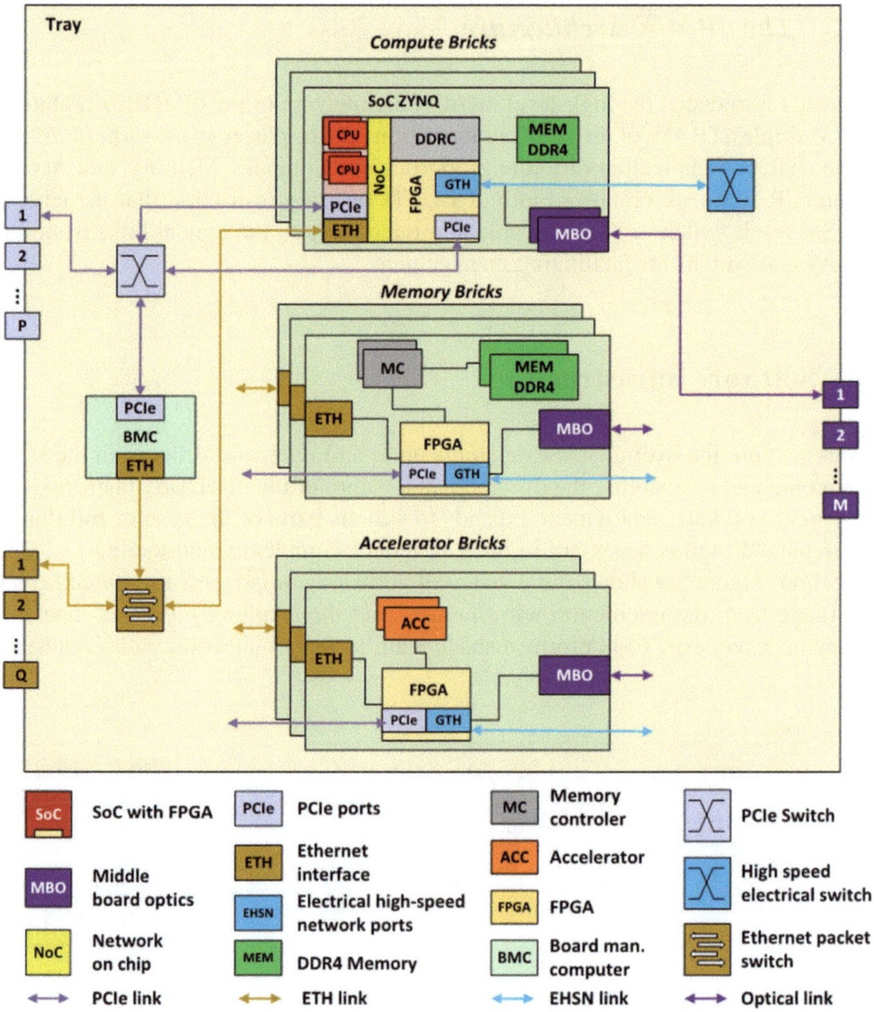

Fig. 3.8 Sample tray architecture with multiple bricks interconnected through optical and electrical interconnection networks

will have a hierarchical topology. Bricks on the same tray will interconnect via a PCIe interface. Inter-brick interconnection between bricks on different trays within the same rack will be provided via a PCIe switch, which will exit the tray with one or more PCIe cable connectors. The PCIe interface will be used for signaling and interrupts, as well as for attachment to remote peripherals. This network can also be used to (re)configure the FPGAs in each SoC.

3.3.3 *The dRACK Architecture*

Figure 3.4 introduced the high-level dRACK architecture of the dReDBox architecture. Multiple dTRAYs of different configurations can be placed in the same dRACK. These dTRAYs can feature different proportions of Compute, Memory, and Accelerators. dRACKs are organized into dCLUSTs due to a restriction that the largest dROSM (rack switch) will not be able to interconnect all the optical links from the dTRAYs, as well as for facilitating management.

3.4 Software Infrastructure

In this section, the overall software architecture and interactions between the software components, enabling the disaggregated nature of the dReDBox platform, are presented. Software deployment expands to various parts of the system and thus a hierarchical design is necessary in order to reduce complexity and facilitate implementation. Figure 3.9 shows the division of software components into three layers that range from the interaction with the user and the handling of virtual machine deployment requests, to platform management, to interconnection paths synthesis

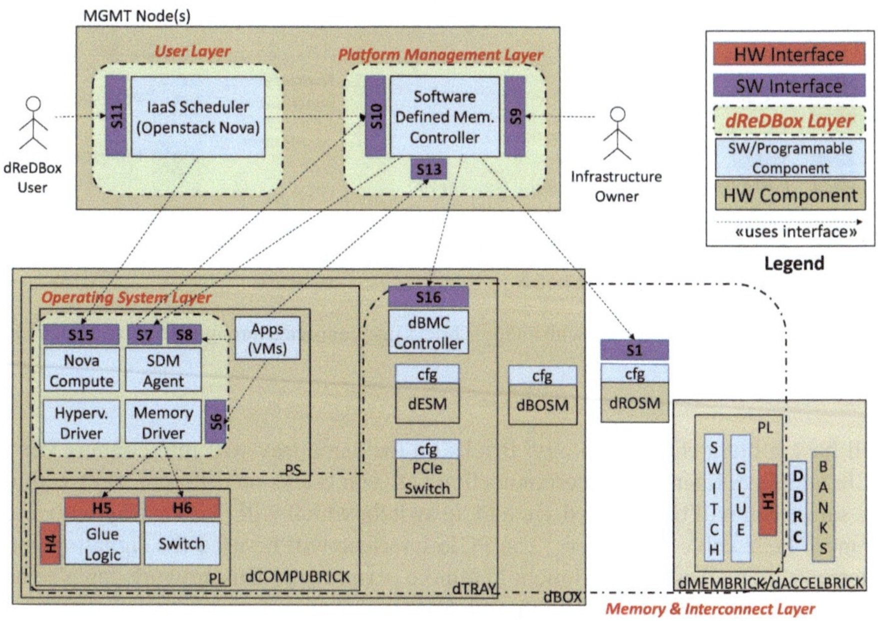

Fig. 3.9 The three dReDBox layers for software support. The user layer provides the end-user interface. The resource management layer enables management and monitoring of resources. The operating system layer on the compute bricks provides hypervisor support to manage virtual machines

and, eventually, to the necessary extensions of the operating system that enables support for remote memory access.

3.4.1 User Layer

This is the layer that serves as the interface between the end user and the system, allowing to reserve virtualized resources and to deploy virtual machines. In addition, it allows the system administrator to register hardware components and monitor them via a graphical user interface.

3.4.1.1 IaaS Scheduler

dReDBox takes advantage of OpenStack [2], the de facto standard cloud operating system, for providing an Infrastructure-as-a-Service platform to provision and manage virtual resources. However, various extensions are required in order to comply with the disaggregated architecture of the dReDBox platform.

OpenStack, through its compute service, Nova [4], employs the scheduler to select the most qualified host in terms of user requirements for compute and memory capacity. This is achieved by filtering and assigning weights to the list of available hosts. In traditional cloud computing systems, resources are provided only by the compute nodes (hosts) and, as a result, scheduling decisions and resource monitoring are limited by this restriction. To take full advantage of disaggregated resources, e.g., memory and accelerators, we modify Nova Scheduler by implementing a communication point with the resource management layer, described below, through a REST API. We, thus, intersect the workflow of virtual machine scheduling and retrieve a suitable compute node for the deployment of a user's virtual machine in order to maximize node's utilization. Furthermore, in the case that the amount of allocated memory to the host does not satisfy user requirements, the resource management layer proceeds by allocating additional remote memory and establishing the required connections between the bricks. Besides virtual machine scheduling, we extend Openstack's web user interface, Horizon [3], in order to enable the administrator to monitor compute bricks, deployed virtual machines, and allocated remote resources. Finally, administrators, through this extended interface, can also register new components (compute bricks, memory modules, etc.) and the physical connections between them.

3.4.2 Resource Management Layer

This is the layer where the management and monitoring of the availability of the various system resources occurs. More specifically, the current state of memory and accelerator allocations is preserved along with the dynamic configurations of the hardware components interconnecting bricks (optical switches, electrical crossbars, etc.).

3.4.2.1 Software-Defined Memory Controller

The core entity of the resource management software is the Software-Defined Memory (SDM) Controller. It runs as a separate and autonomous service, implemented in Python programming language and exposes REST APIs for the interaction with both the IaaS Scheduler and the agents running on compute bricks. Its primary responsibility is to handle allocation requests for both memory and accelerators. Memory requests are arriving to the REST interface from the IaaS scheduler, which is requesting the reservation of resources upon the creation of a virtual machine with a predefined set of requirements for vCPU cores and memory size, called “flavor”. The SDM Controller, then, returns a list of the candidate compute bricks that can satisfy both compute and memory requirements after the consideration of total memory availability in the system and connection points on compute bricks (transceiver ports) and switches (switch ports). Allocation algorithms aiming at improving power savings and/or performance, through low-latency memory accesses, are employed to select the most suitable memory brick for remote memory allocation.

After the selection of the bricks, where the deployment and the memory allocation will take place, a subsequent component, part of the SDM Controller, called platform synthesizer, is employed. Its main function is to maintain an overview of the whole platform hardware deployment and to drive the orchestration and the dynamic generation of system interconnect configurations. From there, these configurations are passed through REST calls to the interfaces managing the electrical crossbars and the optical circuit switches as well as to the agents running on compute bricks, so that they can configure the programmable logic and initiate hotplugging of remote memory segments.

3.4.2.2 Graph Database

The representation of resources and interconnection between them is realized as a directed graph structure implemented in JanusGraph [1] distributed graph database. Resources are modeled as vertices, while the edges between them represent physical connections. Figure 3.10 illustrates a simple example. More specifically, compute/memory/accelerator bricks, as well as transceivers and switch ports are modeled as vertices. We then use directed edges to connect all the transceivers that are located on a compute brick (acting as root) to the transceivers located on memory bricks (acting as sinks). This happens despite the fact that transceivers are bidirectional, but for the sake of convenience we consider transceivers belonging to compute bricks as logical output and transceivers belonging to memory and accelerator bricks as logical input. One or more switching layers exist between transceivers on different bricks. At the first layer, brick (compute/memory/accelerator) transceiver ports are

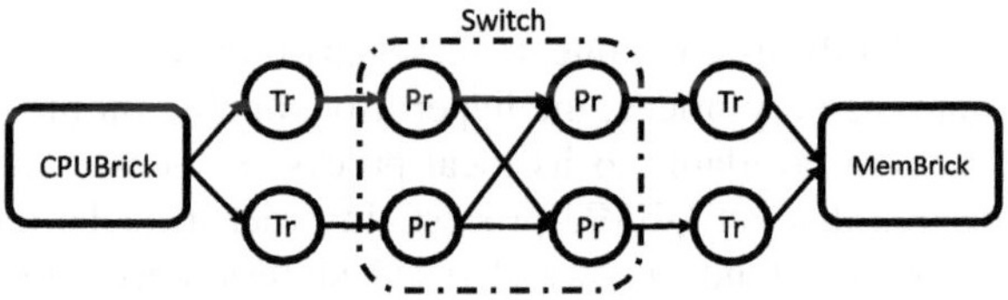

Fig. 3.10 Graph-based resource modeling example for a system that comprises a Compute Brick, a Memory Brick, two transceivers per brick, and one switching layer. Bricks, Transceivers, and Switch ports are depicted as vertices

physically connected either to the ports of the electrical crossbar or to the ports of the optical circuit switch (dBOSM), both of which are residing on the same dBOX as the brick. Remaining ports of the dBOSM are physically connected to the next switching layer, which is the rack switch (dROSM). Switch ports of the same switch are fully interconnected in order to make the establishment of a circuit between any of the switch ports possible. Traversing the edges of the graph that originate at a compute brick and finish at memory bricks allows to retrieve the required paths, i.e., the switch ports that need to be configured for the establishment of the connection between the bricks. After the successful configuration of the components participating in the interconnection, we increase a usage counter defined as a vertex property on each one, denoting that it is already part of the allocation and that it should not be reused for interconnecting additional bricks. Several techniques to avoid loops in traversals and to improve performance have been used, which, however, are beyond the scope of this chapter.

3.4.3 Compute Brick Operating System Layer

The compute brick runs the hypervisor, a modified version of the Linux kernel with KVM modules that add virtual machine management capabilities. In this section, we describe the main operating system level extensions over a traditional Linux system. As expected, innovations in comparison with state-of-the-art systems relate to the management of remote memory and the allocation/deallocation process.

Memory Driver

The memory driver is a collection of hypervisor-level modules supporting dynamic allocation and deallocation of memory resources to the host operating system running on a compute brick. The memory driver implements interfaces which configure the remote memory access, both during guest VM allocation and for dynamic resizing of remote resources. In the rest of this section, we focus on the specification of the main subcomponents of the memory driver, namely, memory hotplug and NUMA extensions.

Once the required hardware components are set up to connect a compute brick with one or more remote memory bricks, the hypervisor running on the compute brick makes the new memory available to its local processes. These processes include VMs, each living in a distinct QEMU process. In order to make remote memory available, the hypervisor extends its own physical address space. This extended part corresponds to the physical addresses that are mapped to remote destinations. Once these addresses are accessed by the local processor, they are intercepted by the programmable logic and forwarded to the appropriate destination.

Memory hotplug is a mechanism that was originally developed to introduce software-side support for server boards that allow to physically plug additional memory SO-DIMMs at runtime. At insertion of new physical memory, a kernel needs to be notified about its existence and subsequently to initialize corresponding new page frames on the additional memory as well as to make them available to new processes. In memory hotplug terminology, this procedure is referred to as a "hot add". Similarly, a "hot remove" procedure is triggered via software to allow detaching pages from a running kernel and to allow the physical removal of memory modules. While originally developed for the x86 architecture, memory hotplug has been ported so far to several different architectures, with our implementation focusing on ARM64. The compute-brick kernel reuses the functionalities provided by memory hotplug to extend the operating system's physical memory space by adding new pages to a running kernel and, similarly, removing them once memory is deallocated from the brick. Unlike the standard use of memory hotplug in traditional systems, there is no physical attachment of new hardware in dReDBox; for this reason, both hot add and hot remove have to be initiated via software in response to remote memory attach/detach requests.

Although the choice of building remote memory attachment on top of Linux memory hotplug allows to save considerable effort by reusing existing code and proven technology, there is a main challenge associated with using it in the context of the proposed architecture. The hotplug mechanism needs to be well integrated with programmable logic reconfiguration, in a way that guarantees that physical addresses as seen by the operating system kernel and content of programmable logic hardware tables are consistent. Non-uniform Memory Access (NUMA) refers to a memory design for single-board multiprocessor systems, where CPUs and memory modules are grouped in nodes. A NUMA node is a logical group of (up to) one CPU and the memory modules which are mounted on the board physically close (local) to the processor. Even though a processor can access the memory on any node of the NUMA system, accessing node-local memory grants significantly better performance in terms of latency and throughput, while performance of memory operations on other nodes depends on the distance of the two nodes involved, which, in traditional systems, reflects both the physical distance on the board and the architecture of the board-level memory interconnect between the processor and the memory module. When a new process is started on a processor, the default memory allocation allocates memory for that process from the local NUMA node. This is based on the assumption that the process will run on the local node and so all memory accesses should happen on the local node in order to avoid the lower latency nodes. This approach works well when dealing with small applications. However, large applications that require more processors or more memory than the local node has to offer will be allocated memory from nonlocal NUMA nodes. With time, memory allocations can become unbalanced, i.e., a process scheduled on a NUMA node could spend most of its memory access-time on nonlocal NUMA nodes. To mitigate this phenomenon, the Linux kernel implements a periodic NUMA balancing routine. NUMA balancing scans the address space of tasks and unmaps the translation to physical address space in order to trap future page faults. When handling a page fault, it detects if pages are properly placed or if they should be migrated to a node local to the CPU where the task is running. NUMA extensions of the Linux Kernel are exploited by our architecture as a means to represent remote memory modules as distinct NUMA nodes: we group remote memory chunks allocated to the brick into one or more CPU-less NUMA nodes. Each of these nodes has its own distance (latency) characterization, reflecting the different latency classes of remote memory allocations (e.g., tray-local, rack-local, or inter-rack). We develop a framework to dynamically provide the Linux Kernel with an overview of the available memory every time new modules are hot plugged to the local system and also provide the distance (latency) between CPUs and allocated remote memory. This information facilitates the development and extension of current task placement and memory allocation techniques within the Linux Kernel for the effective scheduling of VMs to CPUs and for improved locality in memory allocation.

3.5 Simulating Disaggregated Memory

Disaggregated memory is of paramount importance to the approach of disaggregation in data centers, and the capacity to analyze and understand its effect on performance is a prerequisite for the efficient mapping of applications to such architectures. The current section describes a simulation tool for disaggregated memory, dubbed DiMEM Simulator [5, 19] (Disaggregated Memory Simulator), which facilitates the exploration of alternative execution scenarios of typical HPC and cloud applications on shared memory architectures, e.g., Raytrace and Data Caching.

DiMEM Simulator relies on Intel's PIN framework [16] for instrumentation and application analysis purposes, while DRAMSim2 [22] is employed for memory simulations. Figure 3.11 depicts a simplistic view of DiMEM Simulator. On the front end, DiMEM Simulator conducts functional simulations of the TLB and the cache hierarchy, and generates a trace file of accesses to main memory. The *Allcache* pin tool (developed by Klauser [16]) for functional simulations of instruction/data TLB and the cache hierarchy served as the basis in order to support multithreading and hyperthreading. Instrumentation is conducted at the instruction granularity level in order to detect memory accesses that are then analyzed to distinguish between cache hits and misses. The aim is to determine actual accesses to main memory, which inevitably follow the LLC misses. DiMEM Simulator maintains a list of records for every access to main memory, which is used to generate the required memory trace file.

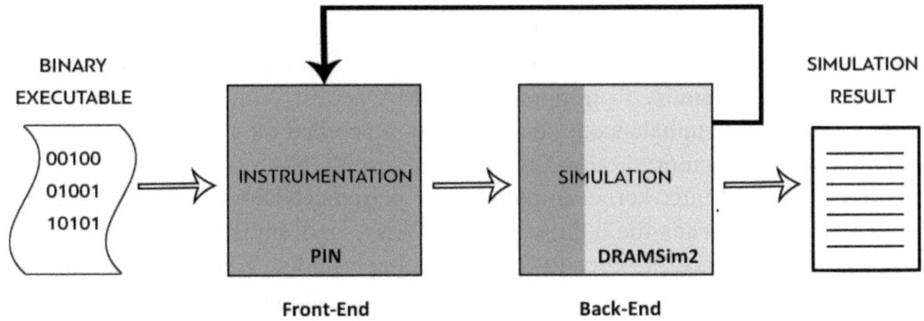

Fig. 3.11 A simplistic illustration of the DiMEM simulation tool for disaggregated memory

On the back end, DRAMSim2 is deployed to analyze the generated memory trace file. A trace file example for DRAMSim2 is shown below, with the first column providing a virtual address, followed by the type of instruction and the cycle it was issued.

```
0x7f64768732d0 P_FETCH 1
0x7ffd16a5a538 P_MEM_WR 8
0x7f6476876a40 P_FETCH 12
0x7f6476a94e70 P_MEM_RD 61
0x7f6476a95000 P_MEM_RD 79
```

Once a certain amount of memory traces is generated, the so-called window, a penalty-based scoring scheme is employed to determine the cycle each instruction is accessing memory. Sorting instructions by the memory access cycle facilitates simulations in multithreaded execution environments. Given the two types of memory that DiMEM is simulating, i.e., local and remote (disaggregated), two DRAMSim2 instances are employed. The second instance is required to model remote memory correctly, given the additional latency of the interconnect.

To evaluate DiMEM Simulator, a variety of HPC and cloud benchmarks were employed. HPC benchmarks involved Barnes, Volrend, and Raytrace of the Splash-3 benchmark suite [23], and FluidAnimate of the PARSEC benchmark suite [6]. The Memcached-based Data Caching benchmark of the CloudSuite [11, 18] benchmark suite for cloud services was employed as the cloud benchmark. All benchmarks were evaluated based on a series of processor configurations that were modeled after modern, high-end microprocessor architectures. Four different memory allocation schemes were employed, with an increasing percentage of remote memory, i.e., 0, 25, 50, and 75%, as well as varying latency for remote accesses, ranging between 500 and 2,000 ns.

Figures 3.12 and 3.13 illustrate Raytrace and Data Caching Workload profiles based on LLC misses measured in Misses per Kilo Instructions (MPKI) per window. As can be observed in the figures, the workload of the Raytrace application exhibits significant variations over time, whereas the Data Caching one remains

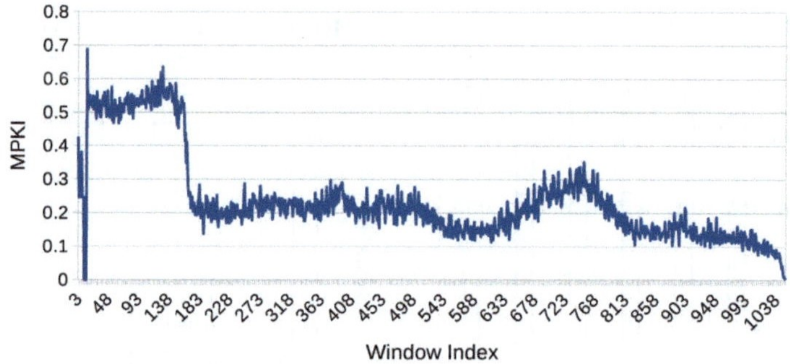

Fig. 3.12 Workload profile for Raytrace (Splash-3 benchmark suite [23]), measured as the number of missed per kilo instruction (MPKI) per window

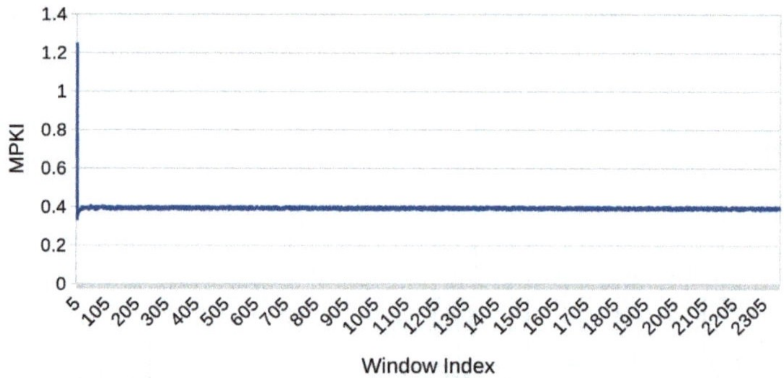

Fig. 3.13 Workload profile for Data Caching (CloudSuite benchmark suite [18]), measured as the number of missed per kilo instruction (MPKI) per window

largely constant. This directs sampling accordingly in order to avoid prohibitively large evaluation times.

Figure 3.14 illustrates the effect of disaggregated memory on application execution in terms of induced overhead/slowdown and varying effective bandwidth as disaggregated latency increases from 500 and up to 2,000 ns. A RISC-based CPU with 4 GB of total memory is employed in all cases, while the percentage of remote memory in the system varies from 25 to 75%. As expected, increasing disaggregated latency reduces the effective bandwidth of the remote memory equivalently in all three remote-memory-usage scenarios, i.e., 25% (Fig. 3.14b), 50% (Fig. 3.14d), and 75% (Fig. 3.14f). As the percentage of remote memory in the system increases, the negative effect of increased remote memory latency on application performance intensifies, as observed in Fig. 3.14a, c, e. As can be observed in the figures, the increased access latency to remote memory reduces overall performance. However, the results suggest that maintaining latency to disaggregated memory below 1,000 ns yields acceptable performance deterioration, allowing to benefit from the overall disaggregated approach.

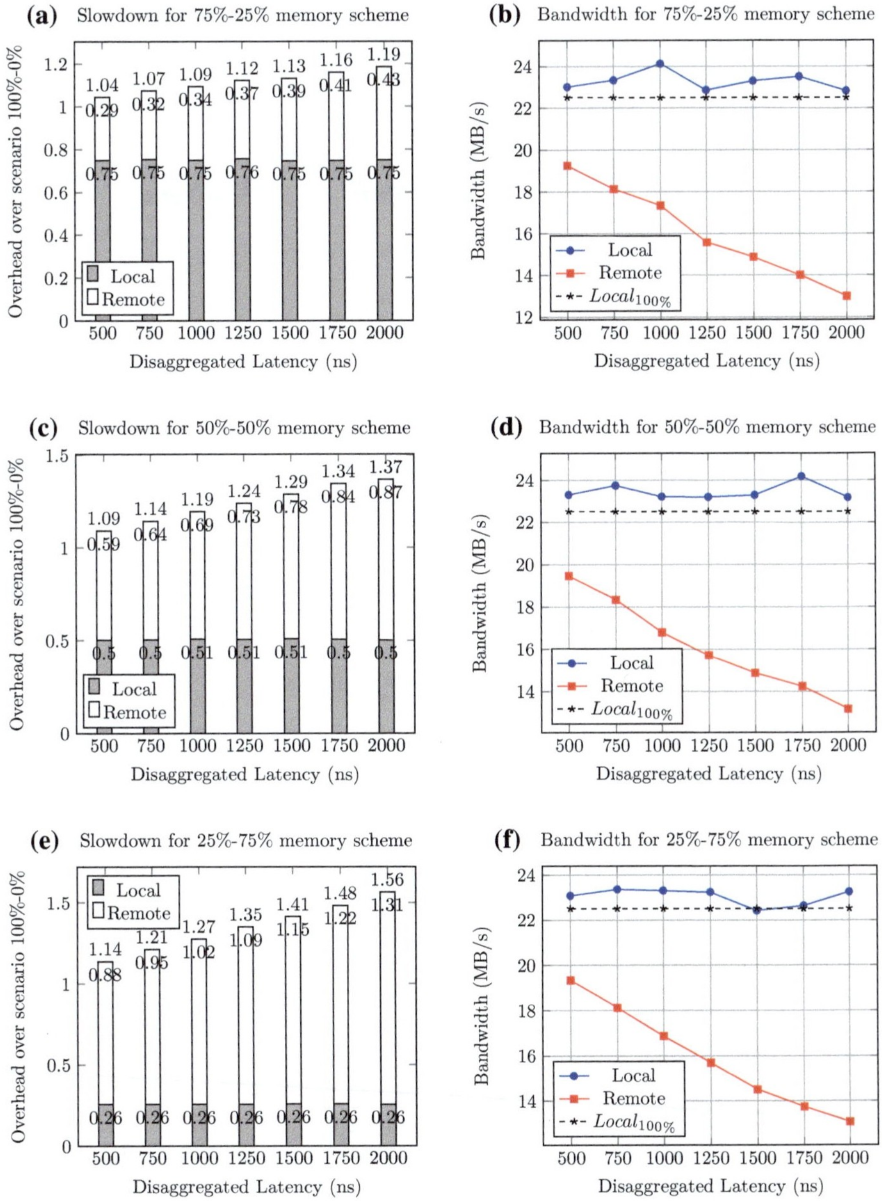

Fig. 3.14 The effect of disaggregated memory on application execution in terms of overhead/slowdown and effective memory bandwidth for varying disaggregated latency values

3.6 Conclusion

The dReDBox project aims at achieving fine-grained resource disaggregation, with function blocks representing the basic units for creating virtual machines. This can lead to fully configurable data center boxes that exhibit the capacity to better serve application requirements, by quantitatively adapting resources to application workload profiles. Compute-intensive applications, for instance, will induce the allocation of increased amounts of CPU nodes, whereas memory-intensive applications will trade processing power for memory and storage resources. Evidently, the success of this approach relies on eliminating disaggregation-induced overheads at all levels of the system design process, in order to virtually reduce the physical distance among resources in terms of latency and bandwidth.

Acknowledgements This work was supported in part by EU H2020 ICT project dRedBox, contract #687632.

References

1. JanusGraph: Distributed graph database (2017). http://janusgraph.org/
2. OpenStack (2017). https://www.openstack.org/
3. OpenStack Horizon (2017). https://docs.openstack.org/developer/horizon/
4. OpenStack Nova (2017). https://docs.openstack.org/developer/nova/
5. Andronikakis A (2017) Memory system evaluation for disaggregated cloud data centers
6. Bienia C, Kumar S, Singh JP, Li K (2008) The parsec benchmark suite: characterization and architectural implications. In: Proceedings of the 17th international conference on parallel architectures and compilation techniques. ACM, pp 72–81
7. Caulfield AM, Chung ES, Putnam A, Angepat H, Fowers J, Haselman M, Heil S, Humphrey M, Kaur P, Kim JY et al (2016) A cloud-scale acceleration architecture. In: 2016 49th annual IEEE/ACM international symposium on microarchitecture (MICRO). IEEE, pp 1–13
8. Chen F, Shan Y, Zhang Y, Wang Y, Franke H, Chang X, Wang K (2014) Enabling FPGAs in the cloud. In: Proceedings of the 11th ACM conference on computing frontiers. ACM, p 3
9. Dragojević A, Narayanan D, Hodson O, Castro M (2014) Farm: fast remote memory. In: Proceedings of the 11th USENIX conference on networked systems design and implementation, pp 401–414
10. Fahmy SA, Vipin K, Shreejith S (2015) Virtualized FPGA accelerators for efficient cloud computing. In: 2015 IEEE 7th international conference on cloud computing technology and science (CloudCom). IEEE, pp 430–435
11. Ferdman M, Adileh A, Kocberber O, Volos S, Alisafaee M, Jevdjic D, Kaynak C, Popescu AD, Ailamaki A, Falsafi B (2012) Clearing the clouds: a study of emerging scale-out workloads on modern hardware. In: ACM SIGPLAN notices, vol 47. ACM, pp 37–48
12. Kachris C, Soudris D, Gaydadjiev G, Nguyen HN, Nikolopoulos DS, Bilas A, Morgan N, Strydis C, Tsalidis C, Balafas J et al (2016) The vineyard approach: versatile, integrated, accelerator-based, heterogeneous data centres. In: International symposium on applied reconfigurable computing. Springer, pp 3–13
13. Klimovic A, Kozyrakis C, Thereska E, John B, Kumar S (2016) Flash storage disaggregation. In: Proceedings of the eleventh European conference on computer systems. ACM, p 29
14. Lim K, Chang J, Mudge T, Ranganathan P, Reinhardt SK, Wenisch TF (2009) Disaggregated memory for expansion and sharing in blade servers. In: ACM SIGARCH computer architecture news, vol 37. ACM, pp 267–278

15. Lim K, Turner Y, Santos JR, AuYoung A, Chang J, Ranganathan P, Wenisch TF (2012) System-level implications of disaggregated memory. In: 2012 IEEE 18th international symposium on high performance computer architecture (HPCA). IEEE, pp 1–12
16. Luk CK, Cohn R, Muth R, Patil H, Klauser A, Lowney G, Wallace S, Reddi VJ, Hazelwood K (2005) Pin: building customized program analysis tools with dynamic instrumentation. ACM SIGPLAN notices vol 40. ACM, pp 190–200
17. Mavroidis I, Papaefstathiou I, Lavagno L, Nikolopoulos DS, Koch D, Goodacre J, Sourdis I, Papaefstathiou V, Coppola M, Palomino M (2016) Ecoscale: reconfigurable computing and runtime system for future exascale systems. In: design, automation & test in Europe conference & exhibition (DATE), 2016. IEEE, pp 696–701
18. Palit T, Shen Y, Ferdman M (2016) Demystifying cloud benchmarking. In: 2016 IEEE international symposium on performance analysis of systems and software (ISPASS), pp 122–132
19. Papadakis O (2017) Memory system evaluation of disaggregated high performance parallel systems
20. Pugsley SH, Jestes J, Balasubramonian R, Srinivasan V, Buyuktosunoglu A, Davis A, Li F (2014) Comparing implementations of near-data computing with in-memory mapreduce workloads. IEEE Micro 34(4):44–52
21. Putnam A, Caulfield AM, Chung ES, Chiou D, Constantinides K, Demme J, Esmaeilzadeh H, Fowers J, Gopal GP, Gray J et al (2014) A reconfigurable fabric for accelerating large-scale datacenter services. In: 2014 ACM/IEEE 41st international symposium on computer architecture (ISCA). IEEE, pp 13–24
22. Rosenfeld P, Cooper-Balis E, Jacob B (2011) Dramsim2: a cycle accurate memory system simulator. IEEE Comput Archit Lett 10(1):16–19
23. Sakalis C, Leonardsson C, Kaxiras S, Ros A (2016) Splash-3: a properly synchronized benchmark suite for contemporary research. In: 2016 IEEE international symposium on performance analysis of systems and software (ISPASS). IEEE, pp 101–111
24. Tu CC, Lee Ct, Chiueh Tc (2014) Marlin: a memory-based rack area network. In: Proceedings of the tenth ACM/IEEE symposium on architectures for networking and communications systems. ACM, pp 125–136
25. Vipin K, Fahmy SA (2014) Dyract: a partial reconfiguration enabled accelerator and test platform. In: 2014 24th international conference on field programmable logic and applications (FPL). IEEE, pp 1–7

Chapter 4
The Green Computing Continuum: The OPERA Perspective

A. Scionti, O. Terzo, P. Ruiu, G. Giordanengo, S. Ciccia, G. Urlini, J. Nider, M. Rapoport, C. Petrie, R. Chamberlain, G. Renaud, D. Tsafrir, I. Yaniv and D. Harryvan

4.1 Introduction

More powerful and smart computing systems are made possible by the continuous advancements in silicon technology. Embedded systems evolved into modern Cyber-Physical Systems (CPS): smart connected systems that are powerful enough to enable (near) real-time interactions with the surrounding environment. For this reason, CPS are at the basis of implementing new services, albeit they generate an enormous amount of data, thanks to their capability of sensing/acting on the environment where they are deployed. Cloud computing, or simply Cloud, is the set of hardware and

A. Scionti (✉) · O. Terzo · P. Ruiu · G. Giordanengo
ISMB, Torino, Italy
e-mail: scionti@ismb.it

O. Terzo
e-mail: terzo@ismb.it

P. Ruiu
e-mail: ruiu@ismb.it

G. Giordanengo
e-mail: giordanengo@ismb.it

S. Ciccia
Politecnico di Torino, Torino, Italy
e-mail: simone.ciccia@polito.it

G. Urlini
STMicroelectronics, Milano, Italy
e-mail: giulio.urlini@st.com

J. Nider · M. Rapoport
IBM Research, Haifa, Israel
e-mail: joeln@il.ibm.com

M. Rapoport
e-mail: rapoport@il.ibm.com

C. Kachris et al. (eds.), *Hardware Accelerators in Data Centers*,
https://doi.org/10.1007/978-3-319-92792-3_4

software technologies used to process and analyse such amount of data, so that it becomes possible to respond to the societal and industrial needs through innovative services. Also, Cloud technologies enables CPS to retrieve useful information to react to the changes in the environment where they operate. However, such welcome capabilities are today counterbalanced by the high power consumption and energy inefficiencies of the processing technologies that traditionally power data center (DC) servers. Generally, DCs have thousands of running servers arranged in multiple racks. Maintaining these infrastructures has high costs, and Cloud providers pay 40–50% of their total expenses as power bills [1]. The critical point of these large-scale infrastructures is represented by their large inefficiency: the average server utilisation remains between 10 and 50% in most of the DCs [2, 3] and further, idle servers can consume up to 65% of their peak power [2, 4, 5]. New architectural solutions are thus required to provide more responsiveness, scalability, and energy efficiency.

One way of ameliorating the situation is to use virtualization techniques [6], which increase server utilisation by replicating many times the behaviour of a physical machine, by means of the abstraction of the main hardware features. Thanks to virtualization, servers with different resources can dynamically host different "virtual machines" (VMs). Hypervisors, i.e. software devoted to control VMs during their lifetime, play a key role in scheduling the allocation of VMs to the physical servers. Also, hypervisors consume computing resources for achieving their goals, which in turn translates into power consumption. In addition, traditional virtualization systems put large pressure on the servers' memory subsystem, thus further contributing to increase power consumption, and limiting the capability of the hypervisors to pack a large number of VMs on the servers. Cloud paradigm allows acquiring and releasing resources dynamically over time, making the problem of allocating computing and storage resources even more complex. The stochastic nature of Cloud workloads, along with blind resource allocation policies (i.e. resources are usually provisioned to their peak usage) employed by most DCs, lead to poor resource utilisation [7, 8].

C. Petrie · R. Chamberlain
Nallatech Ltd, Glasgow, UK
e-mail: c.petrie@nallatech.com

R. Chamberlain
e-mail: r.chamberlain@nallatech.com

G. Renaud
HPE, Grenoble, Israel
e-mail: gallig.renaud@hpe.com

D. Tsafrir · I. Yaniv
Technion–Israel Institute of Technology, Haifa, Israel
e-mail: dan@cs.technion.ac.il

I. Yaniv
e-mail: idanyani@cs.technion.ac.il

D. Harryvan
Certios, Doetinchem, The Netherlands
e-mail: dirk.harryvan@certios.nl

Thus, there is a demand for more efficient resource allocation mechanisms, capable of exploiting the large architectural heterogeneity available in modern DCs, as well as capable of leveraging on less memory-hungry virtualization systems. Furthermore, such allocation strategy should be helpful in reducing the workload imbalance and also to optimise resource utilisation.

Tackling these challenges, the OPERA project [9] aims at investigating on the integration of high-performance, low-power processing elements within highly modular and scalable architectures. OPERA realises that heterogeneity is a key element for achieving new levels of energy efficiency and computational capabilities. To this end, solutions delivered in the project widely leverage on reconfigurable devices (field-programmable gate arrays—FPGAs) and specialised hardware modules to maximise performance and reduce power consumption. Also, improving the efficiency of the virtualization layer represents one of the main objectives of the OPERA project. Improving the way memory is consumed by (virtualized) Cloud applications, adopting more lightweight technologies (e.g., Linux containerisation), and better supporting allocation of heterogeneous computing resources, OPERA aims at greatly reducing the overall energy consumption of next-generation Cloud infrastructures, as well as of remotely connected CPS. Specifically, the following objectives have been identified:

- Exploiting heterogeneity through the adoption of different multicore processor architectures (i.e. ARM, X86_64, and POWER), along with specialised accelerators based on reconfigurable devices (FPGAs);
- Automatically splitting workloads in to a set of independent software modules which can be executed as "microservices" on specific target devices, in order to improve the overall energy efficiency of the Cloud infrastructure;
- Leveraging on direct optical links and the Coherent Accelerator Processor Interface (CAPI) to increase scalability and maximise the performance of the Cloud infrastructure;
- Designing a scalable, modular, and standardised server enclosure (also referred to as "Data Center-in-a-Box") supporting tens to hundreds of processing cores (even with different architectures) in a very compact form factor;
- Defining and integrating appropriate metrics with a more holistic approach to continuously monitor the efficiency of the Cloud computing platform;
- Exploiting ultra-low power multi-/many-cores processors with a dedicated energy-aware instruction set architecture (ISA) to improve CPS processing capabilities in the context of computer vision applications;
- Exploiting reconfiguration and adaptability of the radio communication subsystem to enhance the CPS efficiency;
- Introducing off-loading services (possibly accelerated through FPGA boards) to enable remote CPS accessing larger computing capabilities for critical tasks.

The project, spanning over 3 years, is coordinated by STMicroelectronics, along with the support of Istituto Superiore Mario Boella—ISMB (technical coordinator). The consortium involves also industrial partners, as well as public bodies and academic institutions (HPE, Nallatech, IBM, Teseo–Clemessy, Certios, CSI Piemonte,

Le Départment de Isère, Neavia technologies, Technion) to successfully achieve the above-mentioned objectives.

In this chapter, we discuss the set of technologies, architectural elements, and their integration in the OPERA solution, which are the results of the research activity carried out in the project. Specifically, the design, and integration of a high-density, highly scalable, modular, server equipped with acceleration boards is presented, as well as the capability of integrating, at the DC level, HPC-oriented machines equipped with POWER processors and FPGA accelerators. The integration of an energy-aware Cloud orchestration software is analysed, along with the analysis of the application impact on the memory subsystem, which is at the basis of a more energy-aware optimisation of the running applications. Finally, the extension of the Cloud infrastructure to include a new smart and energy-efficient CPS is presented: thanks to network connections, a mechanism for offloading some of the computation on the Cloud back-end is discussed as well.

The rest of the chapter is organised as follows. Section 4.2 introduces state of the art related works; Sect. 4.3 gives an overview of Cloud paradigm evolution. In Sects. 4.4 and 4.5, we detail OPERA infrastructure resources, spanning from DC servers to remote CPS. Next, in Sect. 4.6, we describe the three real-life application scenarios where OPERA solutions is successfully applied. Finally, we conclude the presented chapter in Sect. 4.7.

4.2 Related Works

The OPERA project exploits many different technologies with the ambition of integrating them into a more (energy) efficient platform, serving as the basis for creating the next-generation Cloud infrastructures.

Improving energy efficiency in Cloud domain has seen several approaches being proposed in recent years [10]. Most of the proposed approaches aim at improving resource allocation strategies, targeting a more efficient use of the computing and storage resource, which, in turn, translates in reducing energy costs. Various algorithms have been used to implement smarter resource allocation strategies, including: best fit [11], first fit [12], constraint satisfaction problem (CSP) [13], control theory [14], game theory [15], genetic algorithms (GA) and their variants [16], queuing model [17] and also various other heuristics [18, 19]. It is interesting to note that greedy algorithms (e.g. best fit and first fit) are employed in commercial products, such as the distributed resource scheduler (DRS) tool from VMware [20]. As an emerging architectural style in developing Cloud applications, splitting them in to a set of independent microservices, it becomes possible to control their allocation on the Cloud resources with a fine-grain resolution with regards to traditional monolithic services, as well as it becomes possible to predict their incoming.

Resource allocation can be conveniently formalised as a Bin Packing Problem (BPP), which is a NP-hard optimisation problem. Generally, solving large instances of BPP requires the adoption of heuristics. Among the various, evolutionary-based algorithms, such as Particle Swarm Optimization (PSO), provide very promising results. Liu and Wang [21] proposed a hybrid PSO algorithm by combining the standard PSO technique with the simulated annealing (SA) to improve the iterations of the PSO algorithm. In [22], the authors introduced a mutation mechanism and a self-adapting inertia weight method to the standard PSO to improve convergence speed and efficiency. Zhang et al. [23] describe a hierarchical PSO-based application scheduling algorithm for Cloud, to minimise both the load imbalance and the inter-network communication cost.

Time series-based prediction has been also used as a mean for foreseeing Cloud workloads in advance, so that it is possible to reserve and optimally allocate DC resources. A comprehensive set of literature is available in this context. A simplified method for modelling univariate time series is given by autoregressive (AR) model. Combined with the moving average (MA), it leads to the well-known ARMA model. In [24], authors proposed a ARMA-based framework for characterising and forecasting the daily access patterns of YouTube videos. Tirado et al. [25] present a workload forecasting AR-based model that captures trends and seasonal patterns. Prediction was used to improve the server utilisation and load imbalance by replicating and consolidating resources. Other works proposed AR-derived models to dynamically optimise the use of server resources and improve energy efficiency [26]. Similarly, the autoregressive integrated moving average (ARIMA) and autoregressive fractionally integrated moving average (ARFIMA) can be used to capture non-stationary behaviours in the time series (i.e. it performs better in describing mean, variance, autocorrelation of time series, which in turn change over time), and to extract statistical self-similar (or long memory) patterns embedded in the data sets. To this end, several research works have been presented [27, 28]. On the other hand, it is also a common practice for the Cloud providers to automatically provisioning virtual machines (VMs) via a dynamic scaling algorithm, without using any predictive method [29, 30].

Heterogeneity (including GPGPUs, FPGAs, Intel XeonPhi, many-cores) have been popularised in HPC domain to accelerate computations and to lower power consumption costs of computing infrastructures. Thanks to different architectural features of the processing elements, heterogeneous solutions allow to better adapt to the workload characteristics. However, they still are not common in the Cloud computing domain [31], where architectural homogeneity helps to reduce the overhead and costs of managing large infrastructures. Although, Cloud providers started offering instances running on powerful GPGPUs (e.g. Amazon AWS G2 and P2 instances, Microsoft Azure N-type instances), the access to FPGA-based images is still complex and very limited, beacause of the difficulties in programming such devices. Some steps towards easing the programming of FPGAs has been done with language extensions and frameworks supporting the translation of code written with popular high-level languages (C/C++) into synthesisable ones (VHDL/Verilog). Examples of such compilation frameworks are represented by LegUp [32], ROCCC [33] and OpenCL [34]. However, the lack of precise control of the placement and routing of FPGA resources, generally still demands for manual optimisation of the final design. Despite many challenges to address, some works tried to reverse the situation, explicitly targeting Cloud workloads [35–37]. However, such works only provide a glimpse of what future evolution of programming languages and compilers will enable (in fact, most of these works strongly rely on an handmade design optimisation phase).

Cloud-connected smart sensors (also referred to as Cyber-Physical Systems) are often demanded for processing intensive tasks to avoid large data transfer to the Cloud back-end. To this end, CPS are equipped with high-performance processors, which run in an ultra-low power envelop. Examples of such design style are PULPino [38], an ultra-low power architecture targeting IoT domain, and the ReISC core architecture [39]. To achieve high-performance within an ultra-low power envelope, CPS processors exploit a mix of energy-optimised technologies, including an energy-aware instruction set architecture (ISA). Despite manufacturing technology and micro-architectural improvements, many application scenarios require computational capabilities (e.g. real-time elaboration of high-resolution streaming videos) that are far from that offered by a CPS. In that case, offloading computational-intensive tasks to a remote Cloud back-end becomes the only feasible solution. RAPID EU-funded project [40] aims at accelerating the capabilities of low-power embedded devices by taking advantage of high-performance accelerators and high-bandwidth networks. Specifically, compute- or storage-intensive tasks can be seamlessly offloaded from the low-power devices to more powerful heterogeneous accelerators. For instance, such mechanism has been successfully used to enable Android devices with no GPU support to run GPU-aware kernels, by migrating their execution on remote servers equipped with high-end GPGPUs [41]. Communication capabilities (both wired and wireless) are at the basis of the implementation of effective offloading services, and remain largely exploited in CPS deployed in unmanned contexts. Adapting the communication subsystem to the physical channel characteristics [42] greatly help the deployment of CPS also in rural areas, as well as greatly contributes to save energy on the battery.

4.3 The Green Computing Continuum

Cloud computing (or simply Cloud) provides a way to access computing resources without hosting them on premise. The level of abstraction at which computing resources are accessed (i.e. the way Cloud users interact with the infrastructure) determines the adopted Cloud service model. Three main service models exist: Software-as-a-Service (SaaS), Platform-as-a-Service (PaaS), and Infrastructure-as-a-Service (IaaS). OPERA aims at improving the way services and resources are managed within a DC, thus it mostly considers the IaaS model. At this level of abstraction (i.e. users may deal with the low-level infrastructure, by requesting virtual machines and managing their interconnections) resource allocation problem emerges as one of the main challenges to address. In addition, recently we witnessed to the broadening of Cloud infrastructures, beyond their traditional concept. With the fast growth of the number of small embedded devices connected to Internet [43], Cloud providers need to support them anywhere at any time. The infrastructural Cloud support helps to

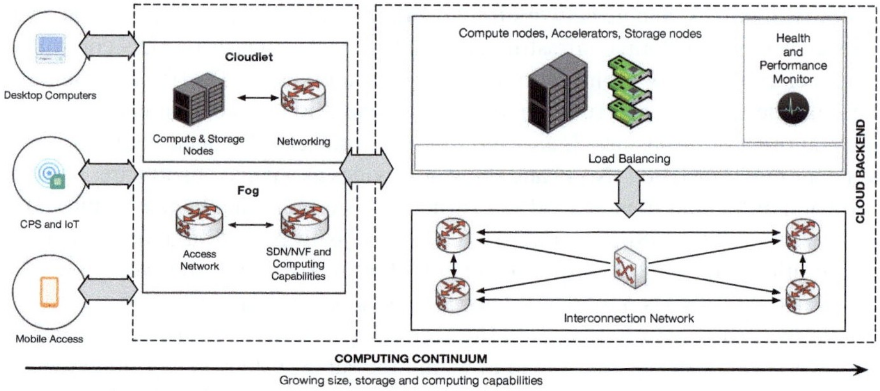

Fig. 4.1 A visual representation of the computing continuum: from left (smart embedded devices) to right (public cloud data centers), computing and storage capabilities grow. Along with compute and storage, energy consumption grows as well

integrate Cloudlets, Fog computing and the "Cloudification" of the Internet of Things (IoT) within traditional infrastructures (see Fig. 4.1). Supporting such growing set of connected devices exacerbates the need for a more efficient way of managing and controlling resources within large-scale computing systems.

Nowadays, mobile computing has gained a momentum thanks to the progress in lowering power consumption of embedded hardware platforms. The higher the computing capabilities are, the more complex the applications running become. Thus, although the improvements in the capabilities of such platforms, Cloud becomes a choice of worth in supporting the remote execution of computing-intensive jobs every time the mobile device is not able to provide enough computational resources. To this purpose, Cloud infrastructures began to reduce the latency for processing jobs and to support (near) real-time services. The adopted solutions, termed as Cloudlets [44], are trusted, capable systems co-located with the point-of-presence, and equipped with a fixed number of powerful servers. Using Cloudlets, users can instantiate, on-demand, custom VMs on a cluster of commodity machines, without the need for accessing traditional Cloud services. Fog computing is another emerging paradigm [45]. Here, the idea is to extend the Cloud infrastructure perimeter, by transferring and processing jobs within the network, aiming at reducing the access latency and improving QoS. Fog computing exploits the "virtualization" of the network equipment: router and switch functionalities are melded with more general-purpose processing capabilities exposed through specific services [46]. The result of such transformation is a set of new technologies, such as software-defined networking (SDN) and network function virtualization (NFV) [47], which are emerging as a front runner to increase network scalability and to optimise infrastructural costs. SDN and NVF are also two examples of Cloud infrastructure elements that can strongly benefit from more efficient, high-density, computing machines (e.g. FPGA devices represents an optimal substrate for accelerating latency-sensitive network operations). Specifically, SDN provides a centralised, programmable network framework that can dynamically provision a virtualized server and storage infrastructure in a DC. SDN consists of SDN controllers (for a centralised view of the overall network), southbound APIs (for communicating with the switches and routers) and northbound APIs (to communicate with the applications and the business logic). NFV helps to run network functions, such as network address translation (NAT) and firewall services as a piece of software. Another extension of the traditional Cloud infrastructures is represented by the Mobile Edge Computing (MEC) [48] paradigm. Here, the aim is to provide a dynamic content optimisation and also to improve the quality of user experience, by leveraging radio access networks (RANs) to reduce latency and increasing bandwidth.

The lowering cost of manufacturing technology led to integrate large computing capabilities, sensors and actuators within the same systems, unleashing the full potential of embedded systems as "cyber-physical systems" (CPS). The large availability of smart sensors/actuators and the corresponding emerging of the Internet of Things (IoT) paradigm [49], makes Cloud infrastructures necessary to store and process the enormous amount of collected data. Cloud-IoT applications are quite different compared to the traditional Cloud applications and services (due to a diverse set of network protocols, and the integration of specialised hardware and software solutions). From this standpoint, Cloud-based IoT platforms are a conglomeration of APIs and machine-to-machine communication protocols, such as REST, WebSockets, and IoT-specific ones (e.g. message queuing telemetry transport—MQTT, constrained application protocol—CoAP).

4.3.1 Energy Efficiency Perspective

Energy efficiency became an important topic for continuing to sustain the adoption of Cloud and IT technologies. Europe, as other countries, defined the objectives that must be achieved to make Cloud and IT technologies "green". To drive energy efficiency in the EU member states, targets have been detailed in the energy efficiency directive, EED [50]. This energy efficiency directive establishes a set of binding measures to help the EU reaching its energy efficiency target by 2020. Under this directive, all EU countries are required to use energy more efficiently at all stages of the energy chain, from its production to its final consumption. Interestingly, energy intensity in EU industry decreased by almost 19% between 2001 and 2011, while the increased use of IT and Cloud services by both individuals as well as organisations has resulted in a marked increase in data center energy use [51]. It is important to note that to decrease final energy use, either total production (output) must decrease and/or efficiency increases must outpace the increase in production.

Although the term "energy efficiency" (E_e) is used often in everyday life, the term warrants careful definition. In mathematical form, all energy efficiency metrics are expressed as

$$E_e = \frac{F_u}{E_{out}} \tag{4.1}$$

where F_u defines the functional unit (i.e. work done), and E_{out} represents the energy used to produce the output results. For the sake of correctly modelling energy efficiency, there are, however, many possibilities for describing (i) energy consumption, (ii) system boundaries, and (iii) functional units (this is mostly represented by the type of workload executed) and test conditions.

Looking at the energy consumption of DC equipment, to get an estimate, one must turn to one or more publically available data sources. These data sources are collected as LCI and LCA databases. LCI databases provide Life Cycle Inventory datasets, while LCA databases include in addition Life Cycle Impact Assessments methods. In environmental impact studies, energy is most often expressed as primary energy (PE). Aside from the grid conversion, LCA studies add other external factors, such as data center cooling, into the PE calculation.

System boundaries are necessary to define in order to correctly calculate the energy efficiency of the system under evaluation. It implies to consider only the elements that actually influence the energy behaviour of the system. Further, workload composition becomes crucial to analyse the efficiency of DC equipment. To this standpoint, three main hierarchical levels can be defined. *System boundary*: this level incorporates all the equipment used in delivery of a service, including end-user equipment and transport networks; *Equipment boundary*: this level is a limitation of the system level, focused on a single machine or a tightly knit cluster of machines delivering certain functionalities (both the user and networking are excluded in this view); *Component boundary*: this level is a limitation of the equipment level focused on an element or component inside a machine or the cluster that performs a distinct function (a component view is not limited to a hardware component but can also be a software component).

Workload greatly influences the behaviour of the hardware units, which in turn determines their consumed energy. To demonstrate the effect of workload, it is useful to examine the results coming from the SPECpower benchmark. Interestingly, low server utilisation is common, even in the case of high-load conditions. Similar outcomes remain valid for the Amazon EC2 service [52], where large fluctuations are present, but the long- term average utilisation is in the range of 10–15%. In this range, even with an aggressive power management, servers do not perform efficiently. Not only workload volume influences the efficiency of computations, but also its composition. Different software components (possibly using different technologies, such as Java virtual machines, compiled languages, etc.) perform differently on a given machine, depending on the other software concurrently running. The influence of workload composition is harder to quantify than that of the workload volume. The issue being that what is measurable in a server is the power drawn along with the application output. Varying the workload composition modifies the application output, making comparison of results impossible. However, such differentiation can be turned into an advantage if heterogeneous hardware elements are used. In this case, it becomes possible to search for an advantage in binding different workload components on different hardware elements, for better performance and energy reduction.

OPERA aims at providing a substantial improvement of the energy efficiency of a DC, by considering workloads made of software components used to deliver end-user services (e.g. access to remote virtual desktops, SaaS-based applications, offloading services, etc.). The influence of the workload volume and composition on the computing elements is also investigated to better tune the architectural design, as well as the adopted technologies. The boundary is represented by the set of machines that perform computations, including also networking equipment and remote CPS (Cloud end-nodes). Thanks to the wide use of heterogeneous platforms, OPERA aims at greatly improving overall energy efficiency.

4.4 Heterogeneous Data Centers

Cloud computing paradigm is based on the availability of a large set of compute, storage and networking resources, which are in general heterogeneous in nature.

Unlike in the past, the growing demand for more energy-aware systems, is leading data centers to rapidly embrace different processor architectures and dedicated accelerators. The former are well represented by ARM-based systems, while the latter are represented by GPGPUs, with a smaller presence of FPGA devices. In OPERA, this practice is further extended by introducing also POWER-based systems. This is a platform originally intended for high-performance computing machines; however, with the emerging of Cloud services supporting HPC-oriented and scientific applications, the availability of dedicated computing nodes becomes more valuable. It is worth to highlight that heterogeneity in data centers extends in several dimensions, not only across architectures: machines configured in the same way and using the same CPU architecture can be still different from each other since CPUs families change the features over generations (e.g. the most recent Intel Xeon processors support the AVX-512 instruction set extension, which was not available on previous models), as well as the supported clock frequencies.

Figure 4.2 depicts the conceptual representation of such modern "heterogeneous" data center, highlighting the presence of software stack devoted to management. Looking at the internal organisation of modern chips, one can see the presence of multiple computing cores, along with dedicated interconnections and accelerating IPs (i.e. circuits providing specific functionalities in hardware). One of the main tasks in governing such kind of infrastructures regards the allocation of resources for the different applications. In order to be "efficient", such allocation should consider several factors to get the optimal allocation decision. To this purpose, in OPERA, power consumption of each platform and the relative load are considered good estimators of the relative energy consumption.

4.4.1 Accelerated Servers

Figure 4.3 depicts the organisation of the heterogeneous DC as envisioned by OPERA. Hardware differentiation provides the proper computational power required

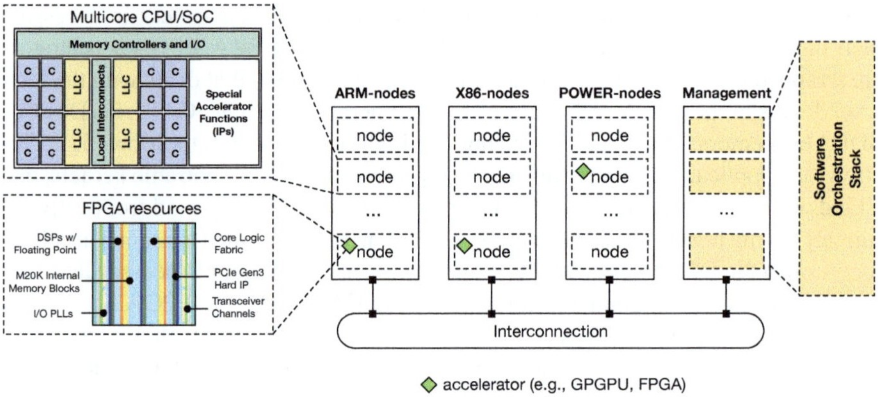

Fig. 4.2 Conceptual representation of a modern heterogeneous data center. Nodes have different instruction sets and micro-architectures, and can be accelerated through GPGPUs or FPGAs. Internal organisation of multicores and FPGAs is provided (left side)

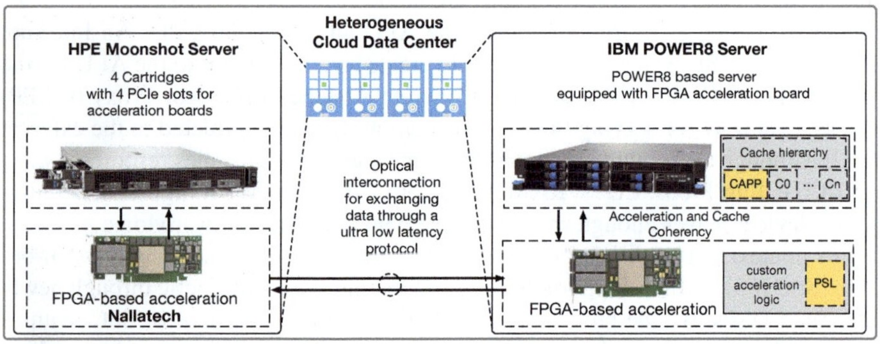

Fig. 4.3 OPERA envisioned heterogeneous Cloud data center infrastructure: high-density, high-performance and low-power server nodes are linked through optical interconnect (via FPGA cards) with POWER-based nodes. FPGA provides also acceleration resources for computational-heavy tasks

by complex tasks running on top of diverse frameworks (e.g. Apache Hadoop, Apache Spark, MPI, etc.).

To this end, OPERA integrates low-power and high-performance processor architectures (ARM, X86_64 and POWER) into highly dense and interconnected server modules. Developed enclosures (HPE Moonshot) makes it possible to integrate hundreds of different processing elements by exploiting a microserver design: a single cartridge contains the specific processing element (i.e. CPU or DSP), the main memory and the storage. Each cartridge can be coupled with a dedicated accelerator (i.e. GPGPU, FPGA-based board, many-core device). Besides Cloud-specific applications, other algorithms may largely benefit from FPGA acceleration, such as network communication functions and protocols.

Effective usage of hardware components depends on the easiness in accessing their functionalities at the software level. To this end, in OPERA, FPGA accelerators are wrapped by an optimised Board Support Package (BSP) that deals with the low-level details of the FPGA architecture and peripherals (e.g. PCIe, CAPI, SerDes I/O, SDRAM memory, etc.), and that fits into the high-level OpenCL toolflow [53]. This allows the application (or a portion) to be represented as highly portable kernels, that Cloud orchestrator can decide at the last minute eventually to off-load from software running on the host processor to silicon gates. Furthermore, such designed accelerators furnish the DC servers with PCIe and Coherent Accelerator Processor Interface (CAPI) attached programmable logic. With the CAPI [54], the FPGA accelerator appears as a coherent CPU peer over the I/O physical interface, thus accessing a homogeneous virtual address space spanning the CPU and the accelerator, as well as a hardware-managed caching system. The advantages are clear: a shorter software path length is required compared to the traditional I/O model. The interface is implemented through two hardware sub-blocks: the Power Service Layer (PSL), and the Accelerator Functional Unit (AFU), i.e. the silicon block implementing the acceleration logic. The PSL block contains the hardware resources that maintain cache coherency, acting as a bridge between the AFU and the main CPU. An Interrupt Source Layer (ISL) is available in order to create an access point to the AFU for the software layer. On the CPU side, the Coherent Attached Processor Proxy (CAPP) acts as a gateway for serving the requests coming from and directed to the external AFU. Although the implementation of PSL and ISL units consumes resources on the FPGA (i.e. Flip-Flops, LUTs, RAM blocks, etc.), the adoption of high-end reconfigurable devices leaves enough space to implement complex hardware logic modules. In this context, OPERA leverages on the last Intel products (Arria 10 System-on-Chip–SoC [55]), which supports IEEE-754 Floating Point arithmetic through newly integrated DSPs. Beside pure reconfigurable logic, the Intel Arria 10 SoC features the second-generation dual-core ARM Cortex-A9 MPCore processor, which is integrated into the hard processor system (HPS) to obtain more performance, flexibility, and security with respect to the previous generation or equivalent soft-cores. On-die ARM cores allow the seamless integration of the reconfigurable logic with a general-purpose elaboration pipeline, and in a broader perspective its integration in the complex DC architecture.

Scalability of the platform is further ensured by the adoption of on-board optical links. OPERA opted for standard Quad Small Form-factor Pluggable (QSFP) modules that permit up to 40 Gb/s of bandwidth, physically configured either as a single 40 Gb/s link or split into four independent 10 Gb/s links. This unprecedented level of flexibility allows for supporting a wide range of topologies. The hardware design issues and constraints are abstracted away and automatically handled by the Intel OpenCL compiler, leaving the software programmer to deal only with specific algorithms of interest. The compiler allows optimising high-level code (e.g. C-based code) enabling OpenCL channels (an OpenCL language construct) to be used for kernel-to-kernel or I/O-to-kernel data transfers at high bandwidth and low latency. Channels are also used to implement an application program interface (API) intended for the host to communicate with the hardware accelerator, generally mapping the PCI Express interconnect.

4.4.2 Workload Decomposition

To take full advantage of hardware specialisation, a mechanism to automatically assign tasks must be put in place. To this purpose, OPERA mostly exploits the *microservice* model. It has recently emerged in the Cloud community as a development style, which allows building applications composed by several small independent but interconnected software modules [56]. Each module runs its own processes and communicates with others by means of a lightweight mechanism, typically consisting of an HTTP-based REST API, resulting in an asynchronous, shared-nothing, highly scalable software architecture. In order to automatise the deployment phase of such microservices-based Cloud applications, an ad-hoc "descriptor" is used. It allows the abstraction of the different software components and their relationships. To this end, OPERA leverages on the OASIS TOSCA [57] standard (Topology and Orchestration Specification for Cloud Applications), which enables the portability of Cloud applications and services across different platforms, and that has recently been extended to support Linux containers. TOSCA provides meta-model expressed with an XML-based language. It consists of two main parts: (i) a topology template—a graph in which typed nodes represent service's components and typed relationships connect nodes following a specific topology; and (ii) plans—workflows used to describe managing concerns (Fig. 4.4).

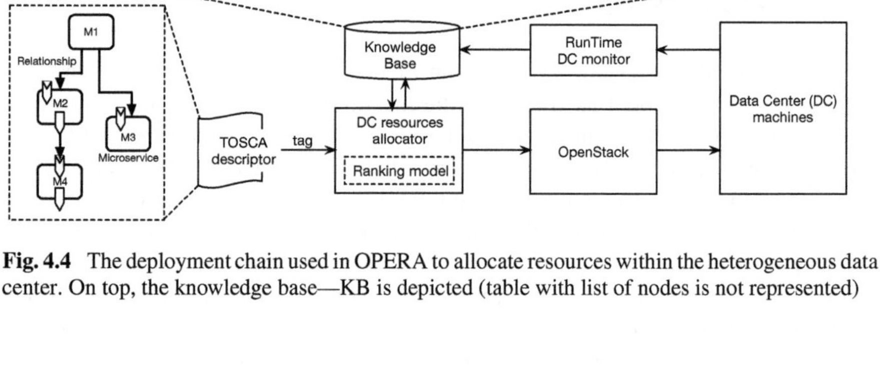

Fig. 4.4 The deployment chain used in OPERA to allocate resources within the heterogeneous data center. On top, the knowledge base—KB is depicted (table with list of nodes is not represented)

In OPERA, the flexibility offered by the TOSCA application descriptor is exploited to create hooks that are used to correctly assign microservices to the most suitable hardware resource for their execution. Although, it is our intention to design and develop a solution that can be used in a different context (i.e. not locked to a specific vendor existing platform), OPERA selected one popular platform as a reference: the OpenStack Cloud orchestration system. With regards to the integration of additional components to the OpenStack system, our solution takes into consideration a 2-steps allocation strategy:

- *Phase–1*: deploy the application components on the most suitable platform (i.e. by choosing the better processor architecture, amount of memory, storage space, etc.), with the aim of maximising the energy efficiency;
- *Phase–2*: periodically rescheduling (i.e. migrating) the application components on the most suitable platform if different load/efficiency conditions arise. For instance, a web front-end previously running on an ARM-based server can be moved on a X86_64 machine if the load of the ARM machine exceeded a threshold and/or the number of requests to the front-end increased.

In this regard, we refer to phase–1 as a *static deployment action*, while we talk of *dynamic (re-)scheduling* of the application components in phase–2. To perform such actions, the Cloud orchestrator needs to match microservices with the most suitable hardware resources based on the indication collected in a Knowledge Base (KB), and to monitor the status of the infrastructure.

Since energy consumption ultimately depends on the power consumption of the host machines in the DC, the enhanced Cloud orchestrator (we can refer to it as the Efficient Cloud Resources Allocation Engine —ECRAE) exploits a simple but rather effective power model to select the host for execution. The power-based model becomes necessary to implement a greedy allocation strategy (Phase–1). Greedy allocation strategy does not ensure an optimal allocation for the whole set of microservices, thus, a further optimisation process is required (Phase–2). Here, a global optimisation algorithm is used to re-schedule all the allocated components with the objective of globally reducing the power (energy) consumption.

4.4.2.1 Efficiency Model

The most critical element in the selection of the actual resource for executing a microservice is the model used to rank the machines belonging to the DC infrastructure. One point to keep into consideration is the relation between energy (E) and power consumption (P). The power P refers to the "instantaneous" energy consumed by a system and generally varies over the time. Based on that, the energy consumed by a system can be computed as the integral of the power consumption function on a given period of time: $E = \int_{t_0}^{t_1} P(t)dt$. The power consumed by a server machine depends on several factors; however, it can be assumed that it is mostly influenced by the consumption of the main hardware components (i.e. CPU, memory, storage and network interface). Power consumption of the CPU and the memory depends on how much the software running on that node stresses these components. Since the load generally changes over time, thus also the power consumed by the CPU and memory (as well as other components) changes. Also, power consumption in idle state is critical. In the literature [2, 5, 58], it has been well documented that a conventional server machine, especially if not properly designed, may consume up to 65% of the maximum power consumption in the idle state. It is also worth noting that given the nature of a service, it becomes difficult to foresee for how much time it will last. For instance, it is not possible to define the amount of time for which a database should run, since it is expected to be accessible unless a failure arises.

To tackle the above-mentioned challenges, we elaborated a simple but still effective model for ranking nodes in the infrastructure. By profiling the behaviour of various microservices, we assume that each software component increases the CPU and memory load for a given quantity. Such quantity (C_l—represents the CPU load increase expressed as a percentage, M_l—represents the memory load increase expressed as a percentage) is measured as the average increase generated by the execution of that software component using the host machine in different working conditions. The following equation allows to emit a score value (R) for a given node:

$$R = \{\alpha \; C_l + (1 - \alpha) \; M_l\} \; P_{avg} \tag{4.2}$$

The score R is the weighted measure of the current power consumption P_{avg} of the node (the power weighted value is biased by the power consumption of the node in idle state, so that the P_{avg} value is given by the power consumption in idle state incremented by the fraction due to the machine load); where the weight is expressed by a linear combination of the current CPU load (C_l) and the memory load (M_l). The α parameter allows to tune the power weight, considering the eventual imbalance between CPU and memory load factors. For instance, setting $\alpha = 0.25$, the load on the memory would be equal to 75%. Power consumption (P_{avg}) is obtained as a measure of the average power consumption of the host platform in different working conditions. Averaging the power consumption allows to capture the typical power profile of the host system. Such value is read directly from the Knowledge Base, thus it must be periodically updated to better reflect real machine behaviour. Such mechanism requires the availability of a hardware power monitor and an interface to query it. As part of the deployment mechanism, the envisioned high-density server enclosures are equipped with such monitoring infrastructure, to ease the action of the ECRAE system.

4.4.2.2 Static Allocation Strategy

TOSCA provides a hierarchical description of a generic Cloud application, which is at the basis of the mechanism used to trigger the “static” allocation strategy. For each element of the hierarchy, the set of scripts to manage the installation and the main functionalities exposed by the component are provided. The last element in the hierarchy is represented by host features. In OPERA, we propose to use a tag to

Algorithm 1: Static allocation strategy

```
Input: Knowledge Base (KB)
Output: DC machine where to execute the microservice
T_c, T_m, α ← get_next_task()
tag ← get_tag()
a_1, a_2 ← get_affinity()
N_1[ ] ← get_node_list(a_1)
N_2[ ] ← get_node_list(a_2)
score[ ] ← Ø
for each n_id in N_1 do
    C_l, M_l, P_avg ← get_node_status(n_id)
    R ← {α C_l + (1 − α) M_l} P_avg
    score[ ] ← add_entry(R, n_id)
score[ ] ← sort(score[ ])
sel_node_id_best_1 ← select_min_score(score[ ], R)
go ← True
while go = True or score[ ] = Ø
do
    sel_node_id_1 ← select_min_score(score[ ], R)
    if sel_node_id_1(C_l) + T_c < 0.98 and
    sel_node_id_1(M_l) + T_m < 0.98 then
        go ← False
    else
        score[ ] ← remove_entry(R, sel_node_id_1)
if go = False then
    sel_node_id_best_1 ← sel_node_id_1
for each n_id in N_2 do
    C_l, M_l, P_avg ← get_node_status(n_id)
    R ← {α C_l + (1 − α) M_l} P_avg
    score[ ] ← add_entry(R, n_id)
score[ ] ← sort(score[ ])
sel_node_id_best_2 ← select_min_score(score[ ], R)
go ← True
while go = True or score[ ] = Ø
do
    sel_node_id_2 ← select_min_score(score[ ], R)
    if sel_node_id_2(C_l) + T_c < 0.98 and
    sel_node_id_2(M_l) + T_m < 0.98 then
        go ← False
    else
        score[ ] ← remove_entry(R, sel_node_id_2)
if go = False then
    sel_node_id_best_2 ← sel_node_id_2
if score[sel_node_id_best_2] < score[sel_node_id_best_1] then
    return sel_node_id_best_2
else
    return sel_node_id_best_1
```

describe the affinity of the software components and the host features. Such affinity is representative of a possible configuration that is evaluated as the most suitable for the execution of the specific component (e.g. a big-memory tag could be used to represent the configuration of a big memory machine, which is well suited for in-memory database operations). The correspondence between the affinity expressed in the TOSCA descriptor and the node configurations is provided by the Knowledge Base. Here, for each configuration, the list of nodes which provide that configuration is also available, along with the actual CPU load, memory load and the average power dissipation. Once the node that better fits with the ECRAE policies has been selected, the corresponding full configuration is used to replace the affinity element in the TOSCA descriptor. The result of such selection process for each application component is a fully compliant TOSCA description file, which can be transformed into a OpenStack compliant description. Interestingly, although the affinity with a specific host configuration is a static information provided in the TOSCA description, OPERA aims at finding and integrating a mechanism able to allow the orchestration system to "automatically learning" which is the best affinity mapping, as a future investigation direction. To the purpose of retrieving information fast, the KB is organised as a (relational) database. Specifically, three tables provide the required information. One table is used to map tags with a list of possible machine configurations. A second table provides the information of each node matching the specific configuration. Finally, a third table provides the current CPU and memory load for each node. This information is updated by an external component (e.g. ceilometer module in the OpenStack, Carbon/Graphite, etc.). All these information then are combined into a simple ranking model (see Eq. 4.2).

The algorithm used by ECRAE is provided as a pseudocode (see Algorithm–1). The first step is to extract the information related to the application component to allocate (lines 1–2). This information regards the increment in terms of CPU and memory loads (as a percentage), the tunable α parameter, and the affinity tag. Given the affinity tag, in line 3 the corresponding configurations (affinities) are extracted. We assume that the first configuration (a_1) represents the best match with the requirements of the application component; however, an alternative configuration can be exploited (a_2). Given the effective configurations, the algorithm extracts the list of nodes in the data center that have those configurations (lines 4–5), then it creates an empty list to associate to each node a corresponding R score. In lines 7–10, a loop is used to create such list (the ranking model described in Sect. 4.4.2.1), and in line 11 the list is sorted. The first element (line 12) is the candidate node providing the lowest power increase, but it is needed to verify if the corresponding increase in the CPU and memory loads are acceptable. To this end, the algorithm analyses all the available nodes (lines 13–21). If none of the nodes can be further loaded (a threshold of 98% is assumed), the algorithm maintains the initial candidate solution. Similarly, in lines 22–40, a candidate solution is searched in the list of alternative configurations. Finally, in lines 41–44, the best ever node is returned. Given this node, the corresponding exact configuration is substituted in the TOSCA descriptor, and the application is allocated through the OpenStack environment (i.e. the Linux container or the required virtual machine are instantiated on the selected node).

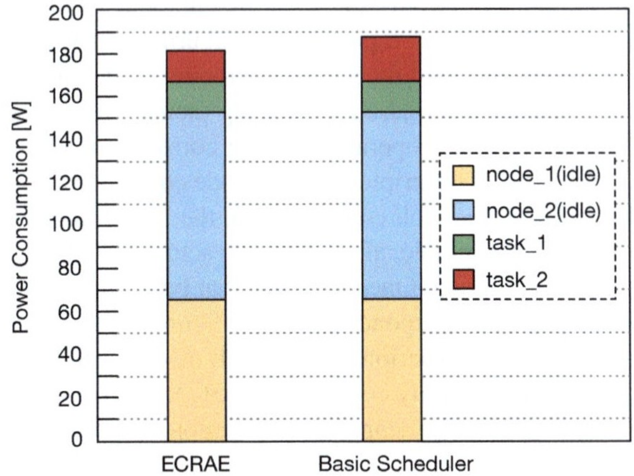

Fig. 4.5 Example of the power (energy) saving provided by the static allocation strategy: two different nodes are considered, respectively, with their idle power consumption

For instance, let us consider two nodes, each of them belonging to one of the flavours associated to a given affinity tag. For the sake of simplicity, we can assume two X86_64 nodes, each in the idle state, but with different average power consumption: we assume $node_1$ consuming up to 100 W (i.e. assuming 65% of idle power consumption that is equal to 65 W), $node_2$ consuming up to 130 W (i.e. assuming 65% of idle power consumption that is equal to 84.5 W). Let us assume to schedule two tasks loading the nodes by 45% each (i.e. the CPU load and memory load are assumed equals to 0.45, using $\alpha = 0.5$). Given this premise, the basic allocation policy (i.e. assigning the task to the less loaded node) leads to a higher power consumption, as reported in Fig. 4.5.

In fact, when the first task is selected, the two nodes are in the idle state and both the strategies (basic and the one implemented by ECRAE) allocate the task to $node_1$. At this point, the power consumption of $node_1$ increases up to 80.75 W, with an overall power consumption, for the two nodes, equals to 165.25 W. On the other hand, the second task is allocated differently. ECRAE ranks the node depending on their weighted power consumption, thus selecting $node_1$ also for the second task (although $node_2$ is less loaded). This provides further 15.75 W of power consumption (increasing the load up to 90%). Conversely, basic allocation strategy selects the node with the lowest load, leading $node_2$ to be selected. In that case, the execution of second task on such node provides 20.475 W of power consumption, leading to an overall power consumption of 185.72 W. Although the power saving for the two nodes is around 5 W, if we consider such saving on a large slice of the DC servers, we obtain a huge improvement in terms of energy saving.

4.4.2.3 Dynamic Allocation Strategy

Static workload allocation exploits an aggressive greedy strategy that does not ensure global optimal allocation of the resources. To guarantee such optimality, an instance of a global scheduling problem (i.e. Bin Packing Problem—BPP) must be solved. Workload scheduling is a well-known optimisation problem that fits in the complexity class of NP-Hard problems. It can be formulated as follows: *Given a set of different objects (VMs or containers), each with a volume S_i (i.e. the amount of CPU and memory used), the objective is to assign as much as possible objects to a bin (i.e. a server machine) that as a finite volume V_j (i.e. a certain amount of CPU and memory offer).* In the context of workload scheduling, the problem requires the minimisation of the number of running machines (i.e. the number of used bins), and also the whole power consumption of the data center (i.e. an heuristic should try to consolidate as much as possible the workload on the minimum number of active hosts). Among various algorithmic solutions, evolutionary-based (and in particular the Particle Swarm Optimisation) techniques allow to quickly solve large instances of this problem.

Particle Swarm Optimisation (PSO) is a population-based stochastic metaheuristic developed by Kennedy and Eberhart in 1995 to optimise multi-modal continuous problems. In PSO, a group of independent solutions (termed as particles) are used to sample the search space and discover the optimal solution. The state of such group of particles is evolved over time, by updating particles' position within the multi-variable search space. Passing from one position in a given instant of time to another is made by taking into account particles' velocity. The velocity and the position of the particles are taken care by two components, which are described as two factors incorporating a form of distributed intelligence:

- *Cognitive factor*: encodes the information regarding the history of the best position assumed by the particles at certain moment in time.
- *Social factor*: encodes the information relating to the history of the best position assumed by the neighbourhood of the particle at certain moment in time.

These two factors are used to adapt the velocity of the particles in such a way it can steer the position towards the optimal solution. PSO does not make use of operators devoted to combining solutions belonging to the same population. On the contrary, the social factor allows to incorporate the knowledge collected by other particles. The topology of the neighbourhood influences the behaviour of the heuristic, although the entire set of particles is used as the neighbourhood (i.e. lattice model). The lattice model also has the advantage of keeping the number of operations used to determine the absolute best position low.

The PSO algorithm is a key component of the ECRAE, since it allows to periodically redistribute the workload on the DC resources, towards their efficient exploitation. At the basis of this periodic imbalance reduction strategy, there is the possibility of migrating application components from one machine to another. Traditional virtual machines can be transparently migrated, but their associated overheads are not acceptable for an effective implementation of the microservices model. To overcome these limitations, Linux containers must be put in place. Linux containers are not designed to be migrated, but such feature becomes important to fully unleash their potential. In the following section, we discuss the way adopted in OPERA to solve this challenge.

4.4.3 Workload Migration

If microservices are in use (i.e. containerised application components), and the microservices have been designed to be scalable and resilient, it is possible that scaling up (i.e. creating more instances) or scaling down (i.e. destroying instances) of a given microservice is the best way to balance the load [59–61]. In other cases, creating or destroying containers is not possible, and we must resort to migration. Virtual machines are more heavyweight example of an execution context, and the price for instantiating new VMs may be considerably higher than migrating an existing one. On the contrary, Linux containers are an example of a lightweight execution context for the microservices model. In the OPERA project, not only we look at how to implement container migration, but also how to perform that in a heterogeneous data center. First, let us list and explain the options that do not make sense. We can then deal with the remaining options as viable, and look at which ones we are focusing on.

In the case of machines with different ISAs (but still general- purpose CPUs) VM migration between two machines may have a merit. Today, it is possible to migrate a VM to a server exposing a different ISA by employing a binary translation mechanism (such as QEMU), which can emulate the target ISA and translate from the source ISA through software functions. This kind of emulation is at least an order of magnitude slower than running native code, which means it should not be used in performance-critical situations. Due to the amount of software running, it also consumes more energy than the native equivalent, and is, therefore, not efficient from several perspectives.

Migrating workloads to accelerators (such as GPGPUs) is a possible avenue that may show results in the future. However, nowadays, due to many restrictions and architectural specialisation, accelerators are not able to manage the execution of a whole VM or container. In fact, such technologies rely on features available on general purpose processors, and they are mostly developed around the widely adopted X86_64 architecture. Although in OPERA the objective is to seamlessly access accelerators (e.g. GPGPUs or even to FPGAs), their usage is limited to the offloading of computational-heavy functions. Here, migration in an heterogeneous context means focusing on multiple ISAs and different types of compute resources, but still relying on general-purpose processors. For these reasons, it does not make sense to migrate a virtual machine or container to such accelerators. It is likely that much more efficient methods of moving a workload to such compute resources will exist. Furthermore, in the case of FPGAs, the acceleration is mainly a static mapping between a computational-heavy function and a dedicated circuit. Even if we are going to consider the case of "high level synthesis" languages such as OpenCL, there is not yet a clear path for migrating general applications (or portions thereof) to such wildly different compute resources. Further, although techniques to dynamically reconfigure the FPGA devices exist and can be used, their overhead remains too high to justify their implementation on a large scale.

4.4.3.1 Container Migration

Container migration (and process migration as the generalised case) is a more relevant problem for which we can develop a solution. Containers gain importance due to the advent of microservices, in which applications are decomposed, and each component is isolated from the others by means of containers. Some containers run services that cannot easily be replicated or scaled (such as in-memory databases), and the cost to migrate such containers is likely to be less than attempting replication. Moreover, container migration is performed at the system level, which means the application does not need to be aware of how to migrate itself, nor that the migration is even taking place. Thus, we may provide migration support for such containerised applications that do not contain support for scaling or replication.

There are several popular implementations of containers on the Linux operating system, such as Docker, LXC/LXD and Runc. Those container implementations rely on Checkpoint-Restore in Userspace (CRIU) tool for checkpointing, restoring and migrating the containers. Although CRIU relies heavily on advanced features found in the Linux kernel, it does not require any modifications of the kernel itself and it is able to perform checkpoint and restore operations entirely in userspace. At the basic level, CRIU allows freezing a running application and checkpointing it as a collection of files. These files can be later used for restoring the application and continuing execution from the exact point where it was frozen. This basic checkpoint-restore functionality enables several features such as application live migration, application snapshots, remote application analysis and remote debugging. Any flavour of Linux containers can be abstracted as a process tree along with additional properties required for process isolation and fine-grained resource management. These processes may have access to various virtual and pseudo devices. CRIU is capable to snapshot the state of the entire process tree as well as the state of the virtual and pseudo devices the processes in the process tree are using. In addition, the properties required for process isolation and fine-grained resource management are saved and become an integral part of the container state snapshot.

4.4.3.2 Comparing Post-copy and Pre-copy Migration techniques

The container state snapshot contains several components that describe process state, open file descriptors, sockets, Linux namespaces, state of virtual and pseudo devices. Yet, all these objects are small and can be easily migrated between different hosts with negligible latency. The part of the container state requiring most of the storage capacity for a snapshot or most of the network bandwidth for a migration is the memory dump of the processes that run in a container. For the case of the container migration, the amount of the memory used by the applications running inside the container defines the time required to migrate the container, as well as the downtime of the application.

The simplest and naive implementation of container migration is as follows: (i) freeze all the processes inside the container; (ii) create a snapshot of the entire container state, including complete memory dump of all the processes; (iii) transfer the container state snapshot to the destination node; and (iv) restore the container from the snapshot. In this case, the time required to migrate the container and the downtime of the application running inside it are equal and both these times are proportional to the amount of memory used by the processes comprising the container. The application downtime during migration may be decreased with one or more round of memory *pre-copy* before freezing the container. With iterative memory pre-copy, container migration time is slightly longer that in the simple case, but the actual downtime of the application is significantly smaller in most cases. However, such approach may not work for applications with rapidly changing memory working set. For such applications, the amount of modified memory will always be higher than the desired threshold, and therefore the iterative pre-copy algorithm will never converge.

An alternative approach for reducing the application downtime is called *post-copy migration* (this is also the approach adopted in OPERA). With post-copy migration, the memory dump is not created and memory contents are transferred after the application is resumed on the destination node. The primary advantage of post-copy migration is its guaranteed convergence. Additionally, post-copy migration requires less network bandwidth than iterative pre-copy migration, since the memory contents are transferred exactly once. The migration time, in this case, is small because only the minimal container state snapshot is transferred before the execution is resumed on the destination node. The application downtime is almost as small as the migration time, however, immediately after migration the application will be less responsive because of the increased latency for memory access (this initial low responsiveness is generally tolerated in Cloud applications).

4.4.4 Optimising Virtual Memory Management

Also, finding the correlation between using different compute resources and runtime help the development of a methodology for performance estimation, in terms of run-time and energy consumption. Accurate and fast methods for performance estimation will be beneficial, for example, in workload management and dynamic allocation of resources. As the number of huge memory pages that can be allocated and the number of threads is limited in increasing performance perspective, then allocating these limited resources between applications that run on the same system requires to have some model to find the best allocation of these resources, to get the best performance. In modern computing platforms and data centers, the DRAM size is not the main performance bottleneck (and energy consumption source), and modern computing platforms can support terabytes of DRAM. But, increasing only the DRAM size does not increase address translation table (TLB), which is used in modern CPUs to quickly access memory locations. Because TLB capacities cannot scale at the same rate as DRAM, TLB misses and address translation can incur crippling performance penalties for large memory workloads. TLB misses might

19. Ghribi C, Hadji M, Zeghlache D (2013) Energy efficient vm scheduling for cloud data centers: exact allocation and migration algorithms. In: 13th IEEE/ACM international symposium on cluster, cloud and grid computing (CCGrid). IEEE, pp 671–678
20. Infrastructure—VMware (2006) Resource management with VMware DRS'. In: VMware Whitepaper
21. Shaobin Z, Hongying H (2012) Improved PSO-based task scheduling algorithm in cloud computing. J Informat Comput Sci 9(13):3821–3829
22. Liu Z, Wang X (2012) A PSO-based algorithm for load balancing in virtual machines of cloud computing environment. In: International conference in swarm intelligence. Springer, pp 142–147
23. Zhang H, Li P, Zhou Z, Yu X (2012) A PSO-based hierarchical resource scheduling strategy on cloud computing. In: International conference on trustworthy computing and services. Springer, pp 325–332
24. Gürsun G, Crovella M, Matta I (2011) Describing and forecasting video access patterns. In: Proceedings of IEEE INFOCOM. IEEE, pp 16–20
25. Tirado, JM, Higuero D, Isaila F, Carretero J (2011) Predictive data grouping and placement for cloud-based elastic server infrastructures. In: Proceedings of the 11th IEEE/ACM international symposium on cluster, cloud and grid computing. IEEE Computer Society, pp 285–294
26. Chandra A, Gong W, Shenoy P (2003) Dynamic resource allocation for shared data centers using online measurements. In: International Workshop on Quality of Service. Springer, pp 381–398
27. Kumar AS, Mazumdar S (2016) Forecasting HPC workload using ARMA models and SSA. In: Proceedings of the 15th IEEE conference on information technology (ICIT). IEEE, pp 1–4
28. Calheiros RN, Masoumi E, Ranjan R, Buyya R (2015) Workload prediction using arima model and its impact on cloud applications' qos. In: IEEE transactions on cloud computing, vol 3, no 4. IEEE, pp 449–458
29. Iqbal W, Dailey MN, Carrera D, Janecek P (2011) Adaptive resource provisioning for read intensive multi-tier applications in the cloud. Fut Generat Comput Syst vol 27, no 6. Elsevier, pp 871–879
30. Beloglazov A, Buyya R (2010) Adaptive threshold-based approach for energy-efficient consolidation of virtual machines in cloud data centers. In: Proceedings of the 8th international workshop on middleware for grids, clouds and e-science ACM. vol 4
31. Crago SP, Walters JP (2015) heterogeneous cloud computing: the way forward. IEEE Comput 48(1):59–61
32. Andrew C, Jongsok C, Mark A, Victor Z, Ahmed K, Tomasz C, Stephen DB, Anderson JH (2013) LegUp: an open-source high-level synthesis tool for FPGA-based processor/accelerator systems. ACM Trans Embed Comput Syst 13(2)
33. Villarreal J, Park A, Najjar W, Halstead R (2010) Designing modular hardware accelerators in C with ROCCC 2.0. In: 18th IEEE annual international symposium on field-programmable custom computing machines. IEEE, pp 127–134
34. Munshi A (2009) The OpenCL specification. In: IEEE hot chips 21 symposium (HCS), pp 1–314
35. Lavasani M, Angepat H, Chiou D (2014) An FPGA-based in-line accelerator for memcached. IEEE Comput Architect Lett 13(2)
36. Putnam A et al (2014) A reconfigurable fabric for accelerating large-scale datacenter services. In: ACM/IEEE 41st international symposium on computer architecture (ISCA). Minneapolis, MN
37. Becher A, Bauer F, Ziener D, Teich J (2014) Energy-aware SQL query acceleration through FPGA-based dynamic partial reconfiguration. In: 2014 24th International Conference on Field Programmable Logic and Applications (FPL), Munich
38. Traber A et al (2016) PULPino: a small single-core RISC-V SoC. In: RISC-V workshop
39. Ickes N et al (2011) A 10 pJ/cycle ultra-low-voltage 32-bit microprocessor system-on-chip. In: Proceedings of the ESSCIRC, Helsinki
40. http://www.rapid-project.eu

41. Montella R, Ferraro C, Kosta S, Pelliccia V, Giunta G (2016) Enabling android-based devices to high-end GPGPUs. In: Algorithms and architectures for parallel processing (ICA3PP)—lecture notes in computer science, vol 10048. Springer
42. Ciccia S et al (2015) Reconfigurable antenna system for wireless applications. In: IEEE 1st international forum on research and technologies for society and industry leveraging a better tomorrow (RTSI), Turin, pp 111–116
43. Evans D (2011) The internet of things how the next evolution of the internet is changing everything. In: CISCO white papers
44. Satyanarayanan M, Bahl P, Caceres R, Davies N (2009) The case for vm-based cloudlets in mobile computing. IEEE Pervas Comput 8(4):14–23
45. Vaquero LM, Rodero-Merino L (2014) Finding your way in the fog: towards a comprehensive definition of fog computing. In: ACM SIGCOMM Computer Communication Review, vol 44, no 5. ACM, pp 27–32
46. Willis DF, Dasgupta A, Banerjee S (2014) Paradrop: a multi-tenant platform for dynamically installed third party services on home gateways. In:Proceedings of the 2014 ACM SIGCOMM workshop on distributed cloud computing. ACM, pp 43–44
47. Martins J, Ahmed M, Raiciu C, Olteanu V, Honda M, Bifulco R, Huici F (2014) ClickOS and the art of network function virtualization. In: Proceedings of the 11th USENIX conference on networked systems design and implementation. USENIX Association, pp 459–473
48. Patel M, Naughton B, Chan C, Sprecher N, Abeta S, Neal A et al (2014) Mobile-edge computing introductory technical white paper. In: Mobile-edge Computing (MEC) industry initiative, white Paper
49. Hwang K, Dongarra J, Fox GC (2013) Distributed and cloud computing: from parallel processing to the internet of things. Morgan Kaufmann
50. European-Commission Energy efficiency directive. https://ec.europa.eu/energy/en/topics/energy-efficiency/energy-efficiency-directive
51. Afman M Energiegebruik Nederlandse commerciele datacenters. http://www.cedelft.eu/publicatie/energy_consumption_of_dutch_commercial_datacentres%2C_2014-2017/1606
52. Huan L Host server CPU utilization in Amazon EC2 cloud. https://huanliu.wordpress.com/2012/02/17/host-server-cpu-utilization-in-amazon-ec2-cloud/
53. Khronos Group The open standard for parallel programming of heterogeneous systems. https://www.khronos.org/opencl/
54. Stuecheli J, Blaner B, Johns CR, Siegel MS (2015) CAPI: a coherent accelerator processor interface. IBM J Res Developm 59(1)
55. Altera Arria 10 FPGAs. https://www.altera.com/products/fpga/arria-series/arria-10/overview.html
56. Thones J (2015) Microservices. IEEE Softw 32(1)
57. Organization for the Advancement of Structured Information Standards (2015) OASIS topology and orchestration specification for cloud applications (TOSCA)
58. Lefurgy C, Wang X, Ware M (2007) Server-level power control. In: Proceedings of the IEEE international conference on autonomic computing. IEEE
59. Roy N, Dubey A, Gokhale A (2011) Efficient autoscaling in the cloud using predictive models for workload forecasting. In: IEEE international conference on cloud computing (CLOUD). IEEE, pp 500–507
60. Chieu TC, Mohindra A, Karve AA, Segal A (2009) Dynamic scaling of web applications in a virtualized cloud computing environment. In: IEEE international conference on e-Business engineering, ICEBE'09. IEEE, pp 281–286
61. Lim HC, Babu S, Chase JS, Parekh SS (2009) Automated control in cloud computing: challenges and opportunities. In: Proceedings of the 1st workshop on automated control for data centers and clouds. ACM, pp 13–18
62. Yaniv I, Dan T (2016) Hash, don't cache (the page table). Sigmetrics

Chapter 5
Energy-Efficient Acceleration of Spark Machine Learning Applications on FPGAs

Christoforos Kachris, Elias Koromilas, Ioannis Stamelos, Georgios Zervakis, Sotirios Xydis and Dimitrios Soudris

5.1 Introduction

Emerging applications like machine learning, graph computations, and generally big data analytics require powerful systems that can process large amounts of data without consuming high power. Furthermore, such emerging applications require fast time-to-market and reduced development times. So to address the large processing requirements of these applications, novel architectures are required in the domain of high-performance and energy-efficient processors.

Relying on Moore's law, CPU technologies have scaled in recent years through packing an increasing number of transistors on chip, leading to higher performance. However, on-chip clock frequencies were unable to follow this upward trend due to strict power-budget constraints. Thus, a few years ago a paradigm shift to multicore processors was adopted as an alternative solution for overcoming the problem. With multicore processors, we could increase server performance without increasing their clock frequency. Unfortunately, this solution was also found not to scale well in the longer term. The performance gains achieved by adding more cores inside a CPU come at the cost of various, rapidly scaling complexities: inter-core communication, memory coherency and, most importantly, power consumption [1].

In the early technology nodes, going from one node to the next allowed for a nearly doubling of the transistor frequency, and, by reducing the voltage, power density remained nearly constant. With the end of Dennard's scaling, going from one node to the next still increases the density of transistors, but their maximum frequency is roughly the same and the voltage does not decrease accordingly. As a result, the power density increases with every new technology node. The biggest

C. Kachris (✉) · E. Koromilas · I. Stamelos · G. Zervakis · S. Xydis
Institute of Communication and Computer Systems (ICCS/NTUA), Athens, Greece
e-mail: kachris@microlab.ntua.gr

D. Soudris
National Technical University of Athens, Athens, Greece

C. Kachris et al. (eds.), *Hardware Accelerators in Data Centers*,
https://doi.org/10.1007/978-3-319-92792-3_5

challenge, therefore, consists of reducing power consumption and energy dissipation per mm^2.

Therefore, the failure of Dennard's scaling, to which the shift to multicore chips is partially a response, may soon limit multicore scaling just as single-core scaling has been curtailed [2]. This issue has been identified in the literature as the *dark silicon* era in which some of the areas in the chip are kept powered down in order to comply with thermal constraints [3]. One way to address this problem is through the utilization of hardware accelerators. Hardware accelerators can be used to offload the processor, increase the total throughput and reduce the energy consumption.

The main contribution of this chapter is a novel framework that is used to extend Spark and allows the seamless utilization of hardware accelerators in order to speedup computational-intensive algorithms in Spark. The proposed framework can be customized, based on the application requirements and allows the utilization of the hardware accelerators with minimum changes in the original Spark code.

The main novelties of the proposed framework are the following:

- An efficient framework for the seamless utilization of hardware accelerators for Spark applications in heterogeneous FPGA-based MPSoCs.
- The development of an efficient set of libraries that hide the accelerator's details to simplify the incorporation of hardware accelerators in Spark.
- Mapping of the accelerated Spark to a heterogeneous 4-nodes cluster of all-programmable MPSoCs (Zynq) based on the Pynq platform.
- A performance evaluation for two usecases on machine learning (Logistic Regression and K-Means) in terms of performance and energy efficiency that shows how the proposed framework could achieve up to 2.5× speedup compared to a high-performance processor and 23× lower energy consumption.
- For embedded applications, the proposed system can achieve up to 36× system speedup compared to embedded processors and 29× better energy efficiency.

5.2 Related Work

In the past few years, there have been several efforts for the efficient deployment of hardware accelerators for cloud computing.

Ghasemi and Chow [4] have introduced a framework for the acceleration of Apache Spark applications. They provide a thorough description of their setup and the bottlenecks that arise as well as experimental results for varying cluster and data sizes. The results are based on a newly implemented accelerated version of the K-Means algorithm. Although their approach is similar to ours, the main difference is found in the programming language that is being used for the implementation of any new APIs. Ghasemi and Chow [4] use JNI to send commands and communicate with the hardware accelerator, while Python cffi objects are used in our case to control the FPGA. Apart from that, their implemented HDFS is consisted of datanodes, that run on each worker while in our case the HDFS runs in pseudo-distributed mode and is hosted on the master node to avoid any additional overhead on the workers side. Finally, we have implemented two hardware accelerators, for Logistic Regression and K-Means algorithms, respectively, and have taken evaluation metrics, proving that our framework is well suited for the ML family algorithms that are compute intensive.

Huang et al. [5] have implemented a framework for the deployment of FGPA accelerators in Spark and Yarn. They provide a generic API to access FPGA resources, while no changes need to be done to Spark or Yarn for the applications to be accelerated and executed on the FPGAs. Although they present a concrete scheme for the resource management of the FPGAs and the tasks scheduling, an additional overhead is added to the dataflow stack, limiting the overall performance. Our work differs as it is based on PySpark and the incorporation of FPGA accelerators in distributed embedded environments through Python libraries.

Chen et al. [6] have implemented a next-generation DNA sequencing application for Spark that uses, PCI-express connected, FPGA accelerators achieving a 2.6× performance speedup. The authors focus on the challenges of such an integration and propose their own solutions while they do not provide a lot of information for the integration of FPGAs to Spark. In this work, we present all the low-level steps we followed including the developed libraries for the communication with the hardware accelerators. Furthermore, we propose a framework that is based on Python and PySpark, making it ideal for the fast development and evaluation of new applications.

Cong et al. [7] present an integrated framework for the efficient utilization of hardware accelerators under the Spark framework. The proposed scheme is based on a cluster-wise accelerator programming model and runtime system, named *Blaze*, that is portable across accelerator platforms. Blaze is mapped to the Spark cluster programming framework. The accelerators are abstracted as subroutines to Spark tasks. These subroutines can be executed either on local accelerators, when they are available, or on the CPU to guarantee application correctness. The proposed scheme has been mapped to a cluster of 8 Xilinx Zynq boards that host two ARM processors and a reconfigurable logic block. The performance evaluation shows that the proposed system can achieve up to 1.44× speedup for the Logistic regression and almost the same throughout for the K-means and 2.32× and 1.55× better energy efficiency, respectively. It has been also mapped to typical FPGA devices connected to the host through the PCI interface. In this case, the performance evaluation shows that the proposed system can achieve up to 3.05× speedup for the Logistic regression and 1.47× speedup for the K-Means and reduces the overall energy consumption by 2.63× and 1.78×, respectively.

There have also been efforts of creating OpenCL kernels for FPGAS including Segal's et al. SparkCL [8] framework that delegates the code to Spark and Aparapi UCores [9] on OpenCL. In other words, Java bytecode is translated to OpenCL creating at runtime any FPGA kernels. Although this approach has potentials, it needs more work to be done and metrics to be taken in order to evaluate the overall benefits in performance and energy consumption and it seems that is more suited to the field of GPUs, where fast compilation times can be achieved.

Kachris et al. [10] present a detailed survey on hardware accelerators for cloud computing applications. In their survey, it is both shown the programming framework that has been developed for the efficient utilization of hardware accelerators as well as the accelerators that have been developed for several applications including machine learning, graph computation, and databases.

IBM has announced in 2016, the availability of SuperVessel cloud, a development framework for the OpenPOWER Foundation. SuperVessel has been developed by IBM Systems Labs and IBM Research based in Beijing. The goal of the SuperVessel cloud is to deliver a virtual environment for the development, testing, and piloting of applications. The SuperVessel cloud framework takes advantage of IBM POWER 8 processors. Developers have access to Xilinx FPGA accelerators, which use IBMs Coherent Accelerator Processor Interface (CAPI) [11]. Using CAPI, an FPGA is able to appear to the POWER 8 processor as if it were part of the processor.

Xilinx has also announced in late 2016, a new framework called *Reconfigurable Acceleration Stack*. This stack is aimed at hyperscale data center that needs to deploy FPGA accelerator. The FPGA boards can be hosted in typical servers and are utilized based on application-specific libraries and framework integration for the five key workloads. These include machine learning inference, SQL query, and data analytics, video transcoding, storage compression, and network acceleration [12]. According to Xilinx, the acceleration stack based on the FPGAs can deliver up to 20× acceleration over traditional CPUs with a flexible, reprogrammable platform for rapidly evolving workloads and algorithms.

Byma et al. [13] introduce a novel approach for integrating virtualized FPGA-based hardware resources into cloud computing systems with minimal overhead. The proposed framework allows cloud users to load and utilize hardware accelerators across multiple FPGAs using the same methods as the utilization of Virtual Machines. The reconfigurable resources of the FPGA are offered to the users as a generic cloud resource through OpenStack.

In this chapter, we give a thorough description of the framework both in terms of software and hardware components and an extended performance evaluation of two widely used machine learning algorithms. Specifically, in this chapter, we present a seamless utilization of hardware accelerators that can be used both for embedded systems and high-performance applications that are based on the Spark framework for computational-intensive applications like machine learning and graph computation. The proposed framework allows the transparent utilization of the hardware accelerators based on the Spark framework using the accelerators as typical python packages.

5.3 Apache Spark

One of the typical applications that are hosted in cloud computing is data analytics. Apache Spark [14] is one of the most widely used frameworks for data analytics. Spark has been adopted widely in recent years for big data analysis by providing a fault-tolerant, scalable, and easy to use in-memory abstraction. Specifically, Spark provides programmers with an application programming interface centered on a data structure called the resilient-distributed dataset (RDD). RDD is a read-only multiset of data items distributed over a cluster of machines, that is maintained in a fault-tolerant way [15]. It was developed in response to limitations in the MapReduce cluster computing framework, which forces a particular linear dataflow structure on distributed programs. MapReduce programs read input data from disk, map a function across the data, reduce the results of the map, and store reduction results on disk. Spark's RDDs function as a working set for distributed programs that offers restricted form of distributed shared memory. Therefore, the latency of such applications, compared to Apache Hadoop, may be reduced by several orders of magnitude.

When the user runs an *action* (like collect), a *Graph* is created and submitted to a DAG Scheduler. The DAG scheduler divides operator graph into (map and reduce) stages. A stage is comprised of tasks based on partitions of the input data. The DAG scheduler pipelines operate together to optimize the graph. The final result of a DAG scheduler is a set of stages. The stages are passed on to the Task Scheduler. The task scheduler launches tasks via cluster manager. The *Worker* then executes the tasks for the task processing [15], as is depicted in Fig. 5.1.

Spark libraries cover four main categories of applications: machine learning (MLib), graph computations (GraphX), SQL query, and streaming applications.

- Spark MLlib is a scalable machine learning library consisting of common learning algorithms and utilities, including classification, regression, clustering, collaborative filtering, dimensionality reduction, as well as underlying optimization primitives.
- GraphX is a Spark API (Application Programming Interface) for graphs and graph-parallel computation. At a high level, GraphX extends the Spark RDD by introducing the Resilient- Distributed Property Graph: a directed multi-graph with prop-

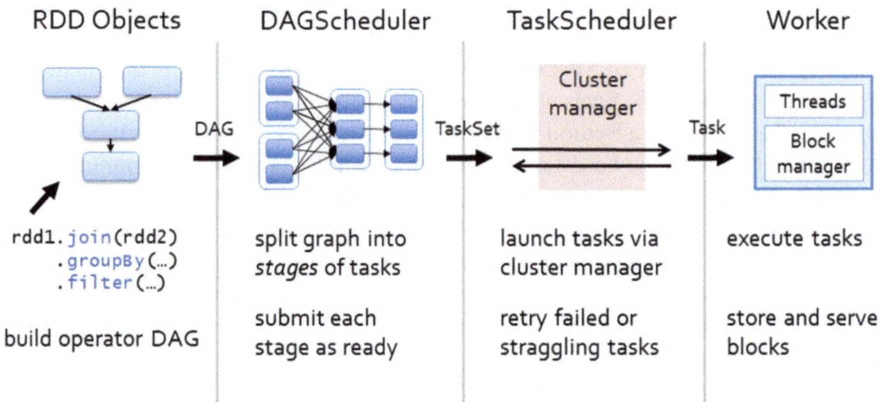

Fig. 5.1 The Spark framework

erties attached to each vertex and edge. GraphX includes a growing collection of graph algorithms and builders to simplify graph analytics tasks.
- Spark SQL provides the capability to expose the Spark datasets over JDBC API and allow running the SQL-like queries on Spark data using traditional business intelligence (BI) and visualization tools.
- Spark Streaming can be used for processing the real-time streaming data. This is based on micro-batch style of computing and processing. It uses the DStream, which is basically a series of RDDs, to process the real-time data.

5.4 SPynq: A Framework for Spark Execution on a Pynq Cluster

PYNQ [16] is an open-source project from Xilinx that makes it easy to design embedded systems with Xilinx Zynq AP SoCs. Using the Python language and libraries, designers can exploit the benefits of programmable logic and microprocessors in Zynq to build more capable and exciting embedded systems. PL circuits are presented as hardware libraries called *overlays*. These overlays are analogous to software libraries. A software engineer, can select the overlay that best matches their application and access it easily through an Application Programming Interface (API). The PYNQ-Z1 board is the hardware platform for the PYNQ open-source framework and is based on the Zynq all-programmable SoC. On top of the Pynq framework, we have efficiently mapped the Spark framework and we have adapted it to communicate with the hardware accelerators located in the programmable logic of the Zynq system. Spark master node is hosted on a personal computer that comes with an Intel i5 ×86_64 architecture processor, but also an Intel ×86 or ARM system could be used. Worker nodes are hosted on PYNQ-Z1's ARM cores. Figure 5.2 shows the proposed cluster architecture.

In addition, Spark comes with three different cluster managers: Standalone, Yarn, and Mesos. For the specific evaluation, the standalone manager is used in client mode, meaning that the driver of Spark is launched in the same process as the client that submits the application. Each worker node is configured for starting one executor instance. Furthermore, a Python API is used for each accelerator that is used for the communication with the hardware accelerator. Each Python API is communicating with the C library that serves as the hardware accelerator driver.

On the reconfigurable logic part, the hardware accelerators for the specific application are hosted. The hardware accelerators are invoked by the Python API of the Spark application. Therefore, the only modification that is required is the extension of the Python library with the new function calls for the communication with the hardware accelerator.

In the typical case, the Spark application invokes the Spark MLlib and this library utilizes the Breeze library (a numerical processing library for Scala). Breeze library

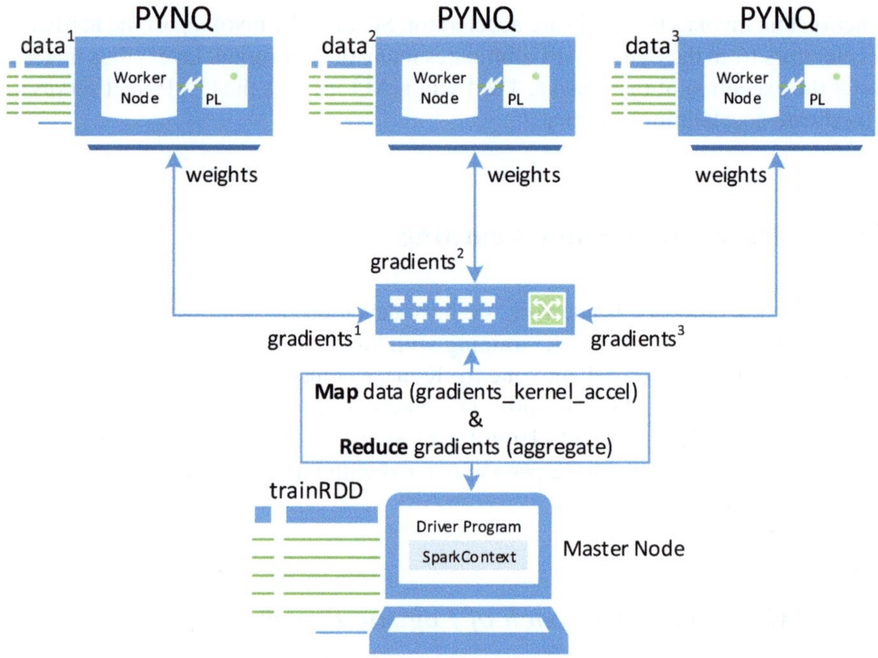

Fig. 5.2 Spynq architecture (logistic regression mapreduce)

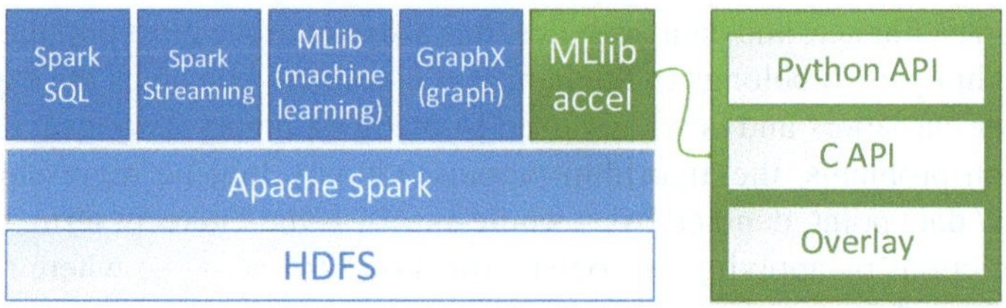

Fig. 5.3 SPynq ecosystem and its layers of abstraction

invokes the Netlib Java framework that is a wrapper for low-level linear algebra tools implemented in C or Fortran. Netlib Java is executed through the Java Virtual Machine (JVM) and the actual linear algebra tools (BLAS—Basic Linear Algebra Subprograms) are executed through the Java Native Interface (JNI).

All these layers add significant overhead to the Spark applications. Especially in applications like machine learning, where heavy computations are required, these layers add significant overhead to the computational kernels. Most of the clock cycles are wasted for passing through all these layers (Fig. 5.3).

The utilization of hardware accelerators directly from Spark has two major advantages; first, the application in Spark remains as it is and the only modification that is required is the replacement of the machine learning library's function with the function that invokes the hardware accelerator. Second, the invoking of the hardware accelerators from the Python API eliminates many of the original layers thus making faster the execution of these tasks. The Python API invokes the C API that serves as a hardware acceleration's library.

5.5 Usecases on Machine Learning

To evaluate the proposed framework, we have developed two hardware accelerators, one for Logistic Regression (LR) training with Gradient Descent and more specifically for the gradients kernel, and one for K-means clustering and more specifically for the computation of the centroids. The hardware accelerators have been implemented using the Xilinx Vivado High-Level Synthesis (HLS) tool. The algorithms have been written in C and have been annotated with HLS *pragmas* for the efficient mapping in reconfigurable logic.

5.5.1 Algorithmic Approach of Logistic Regression

Logistic Regression is used for building predictive models for many complex pattern-matching and classification problems. It is used widely in such diverse areas as bioinformatics, finance, and data analytics. It is also one of the most popular machine learning techniques. It belongs to the family of classifiers known as the exponential or log-linear classifiers and is widely used to predict a binary response. For binary classification problems, the algorithm outputs a binary logistic regression model. Given a new data point, denoted by x, where $x_0 = 1$ is the intercept term, the model makes predictions by applying the logistic function $h(z) = \frac{1}{1+e^{-z}}$, where $z = w^T x$.

By default, if $h(w^T x) > 0.5$, the outcome is positive, or negative otherwise, though unlike linear SVMs (Support Vector Machines), the raw output of the logistic regression model, $h(z)$, has a probabilistic interpretation (i.e., the probability that x is positive).

Given a training set with $numExamples$ (n) data points and $numFeatures$ (m) features (not counting the intercept term) $\{(x^0, y^0), (x^1, y^1), (x^{n-1}, y^{n-1})\}$, where y^i is the binary label for input data x^i indicating whether it belongs to the class or not, logistic regression tries to find the parameter argument w (weights) that minimizes the following cost function:

$$J(w) = -\frac{1}{n}\sum_{i=0}^{n-1}\{y^i \log[h(w^T x^i)] + (1 - y^i)\log[1 - h(w^T x^i)]\}$$

The problem is solved using Gradient Descent (GD) over the training set (α is the learning rate):

```
procedure train(x, y)
  initialize w with zero
  while not converged:
    gradients_kernel(x, y, w)
    for every j = 0, ..., m − 1:
      w_j −= (α/n) g_j

procedure gradients_kernel(x, y, w)
  for every j = 0, ..., m − 1:
    g_j = Σ_{i=0}^{n−1} {[h(w^T x^i) − y^i] x_j^i}
```

For multi-class classification problems, the algorithm compares every class with all the remaining classes (One vs. Rest) and outputs a multinomial logistic regression model, which contains $numClasses$ (k) binary logistic regression models. Given a new data point, k models will be run, and the class with largest probability will be chosen as the predicted class.

5.5.2 Algorithmic Approach of K-Means

K-means is one of the simplest unsupervised learning algorithms that solve the well-known clustering problem and is applicable in a variety of disciplines, such as computer vision, biology, and economics. It attempts to group individuals in a population together by similarity, but not driven by a specific purpose.

The procedure follows a simple and easy way to cluster the training data points into a predefined number of clusters (K). The main idea is to define K centroids c, one for each cluster.

Given a set of $numExamples$ (n) observations $\{x^0, x^1, x^{n-1}\}$, where each observation is an m-dimensional real vector, K-means clustering aims to partition the n observations into K $(\leq n)$ sets $\{s^0, s^1, s^{K-1}\}$ so as to minimize total intra-cluster variance, or, the squared error function:

$$J = \sum_{k=1}^{K} \sum_{x \in s^k} ||x - c^k||^2$$

The K-means clustering algorithm is as follows:

procedure $train(x)$
  initialize c with K random data points
  while not converged:
    $centroids\_kernel(x, c)$
    for every $k = 0, ..., K - 1$:
      $c^k = \frac{1}{|s^k|} \sum_{x \in s^k} x$

procedure $centroids\_kernel(x, c)$
  for every $k = 0, ..., K - 1$:
    $s^k = \left\{x : \left\|x - c^k\right\|^2 \leq \left\|x - c^{k'}\right\|^2 \forall k', 0 \leq k' \leq K - 1\right\}$

The algorithm as described, starts with a random set of K centroids (c). During each update step, all observations x are assigned to their nearest centroid, while afterward, these center points are repositioned by calculating the mean of the assigned observations to the respective centroids.

5.5.3 *Hardware Implementation*

Known techniques, including source-to-source optimizations (like loop transformations) and memory partitioning, were used to implement hardware designs that exploit the inherent data parallelism of these ML formulas, on a macroscopic and microscopic level. The first one is based on advancing coarse-level parallelism and relies on the idea of executing multiple instances of the computational kernels at the same time, each instance operating on a subset of the initial set. The second technique includes implementing loop transformations to the code manually and selecting HLS directives to exploit the data parallelism of the algorithms. A critical step to create the optimal hardware design for a specific function, is to perform design space exploration, with the available hardware resources, in terms of DSPs, BRAMs, etc., being the limiting factor.

Figure 5.4 depicts a minimalistic block diagram of the Logistic Regression accelerator. We would like to note here that AXI4-Stream Accelerator Adapter IP is also used as an intermediate module (while it does not appear in the figure), in order to control the variable size of the data chunks. Other necessary IP blocks, like AXI4-Stream Data Width Converter, etc., are also used, but are omitted from the diagram for simplicity reasons. Four different AXI streaming channels are used for the communication between the ARM cores and the hardware accelerator; two channels are used for sending the data and one channel is used for sending the weights. One

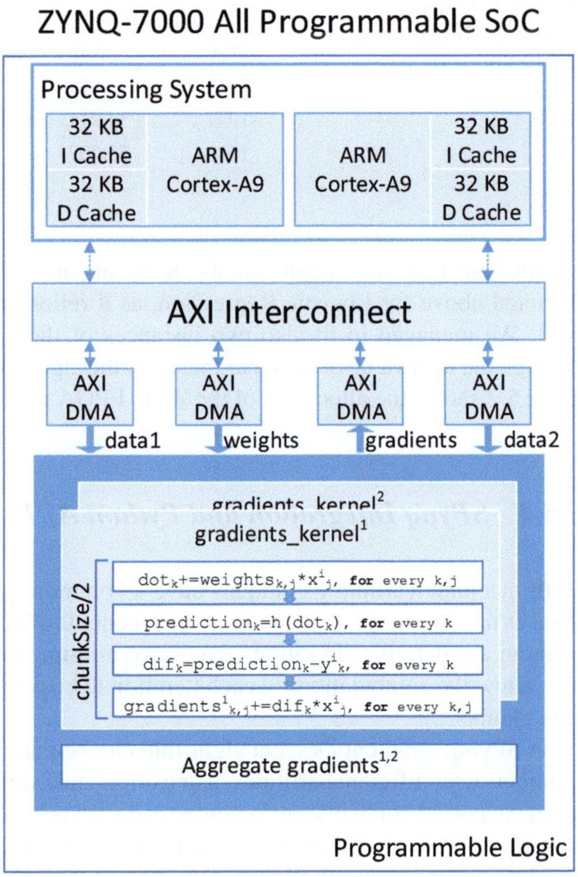

Fig. 5.4 Acceleration of logistic regression using Zynq-7000 AP SoC

Table 5.1 Resource allocation of the logistic regression accelerator

Resources	Used	Total	Utilization (%)
DSP	160	220	73
BRAM	73	140	52
LUT	40,612	53,200	76
FF	60,852	106,400	57

more channel is used to receive the results of the accelerator (gradients). Finally, to speedup the execution time, the programmable logic hosts two instances of the kernel (*gradients_kernel*) that can be running in parallel. Each instance consists of four loop nests corresponding to the four operations in the gradient formula.

In terms of resource allocation, Table 5.1 shows the utilization of the hardware resources in the Zynq FPGA, for Logistic Regression accelerator.

Table 5.2 Resource allocation of the K-means accelerator

Resources	Used	Total	Utilization (%)
DSP	142	220	65
BRAM	135	140	96
LUT	34,911	53,200	66
FF	52,207	106,400	49

The interface architecture of the K-means accelerator is similar with the one depicted above for Logistic Regression, as it relies on AXI Streaming protocol as well. We managed to fit also two instances of the *centroids_kernel*, restricting through the max value of the k parameter of the algorithm, due to BRAM limitations. Table 5.2 shows the allocation of the Zynq FPGA resources, for this accelerator.

5.5.4 SPynq Integration and Python API

Both machine learning techniques have a common characteristic that makes them ideal as means to explore the performance benefits of our heterogeneous cluster. Both of them are iterative algorithms that make multiple passes over the data set, while also allow the computations in each iteration to be performed in parallel on different data chunks.

In SPynq, Gradient Descent algorithm can be parallelized by averaging the sub-gradients over different partitions, using one standard Spark MapReduce in each iteration. So partial gradients are computed in each worker (exploiting the Programmable Logic), using different chunks of the training set, and then Master aggregates them and updates weights. Similarly, in K-means, the computation of the partial sums and counts for each new cluster is performed on the available Workers, and then the Master aggregates the results and calculates the new centroids.

Taking into account and understanding the structure and the nomenclature of Sparks MLlib, we developed new libraries for Logistic Regression with Gradient Descent and K-means clustering, that take advantage of the PL that is available in the PYNQ workers. As a result, when a Spark user wants to utilize the hardware accelerator in an existing application, the main change that needs to be made, is the replacement of Sparks mllib library, that is imported, with our mllib_accel one, as shown in the following code example. Therefore, a user can speedup the execution time of a Spark application by simply replacing the library package.

```
from pyspark import SparkContext
from mllib.regression import LabeledPoint
from mllib_accel.classification import LogisticRegression

[...]

sc = SparkContext(appName = "Python LR")

trainRDD = sc.textFile(train_file, numPartitions).map(parsePoint)
testRDD = sc.textFile(test_file, numPartitions).map(parsePoint)

LR = LogisticRegression(numClasses, numFeatures) · train(trainRDD, chunkSize,
      alpha, numIterations)
LR.test(testRDD)

sc.stop()
```

The first approach was to simply replace the mapper functions (*gradients_kernel* and *centroids_kernel*) with Python APIs that drive the hardware accelerators. Inside these new functions we used to download the equivalent overlay, create the necessary DMA objects, store the data inside the corresponding buffers, perform the DMA transfers and finally destroy them, free the allocated memory and return the results. After profiling the applications, we concluded that most of the time (99%) is wasted on writing the train RDD data to the allocated DMAs buffers.

However, since the data remain the same (cached) over the whole execution of the training, we have managed and implemented a novel scheme that allows the persistent storing of the RDD in contiguous memory, avoiding in-memory transfers every time the accelerator is invoked. For this reason, we developed a new mapper function that allocates and fills contiguous memory buffers with the training data, in order to remain there for the rest of the application execution. So when the DMA objects are created in each iteration, there is no need to create new buffers for them and fill them with the corresponding data, they just get assigned the previously created ones. Also, before destructing these DMA objects, their assigned buffers are set to 'None', so that they remain intact and are not freed as it is shown in Fig. 5.5.

Based on the above, we have created Python APIs which basically consist of three calls:

- **CMA** (Contiguous Memory Allocate): This call is used for the creation of the buffers and the further allocation of contiguous memory. Also at this point, the overlay is downloaded and the training data are written to the corresponding buffers. Using cma, a new RDD, which contains only information about these buffers (memory addresses, sizes, etc.), is created and persisted.
- **kernel_accel** (gradients/centroids): In this call, the DMA objects are created using Xilinxs built-in modules and classes; previously allocated buffers are assigned to DMAs, current weights/centers are written in memory and finally data are

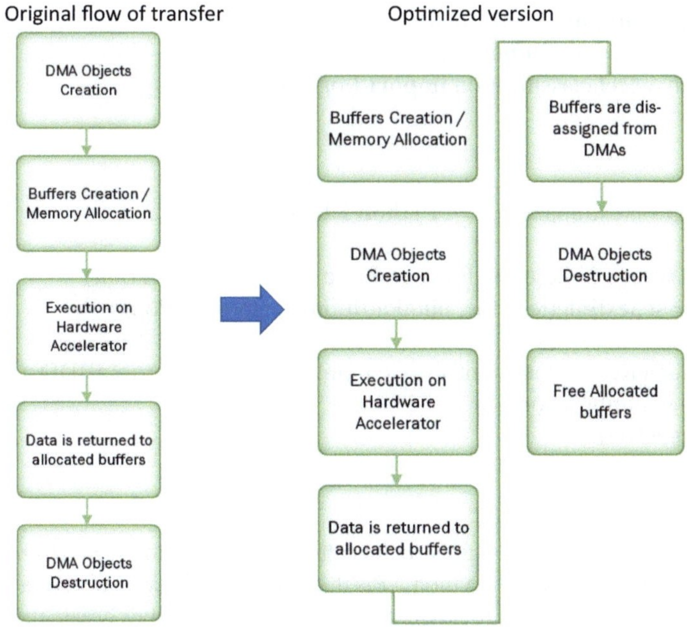

Fig. 5.5 Flow of the original and optimized method for the DMA transfers to the accelerator

transferred to the programmable logic. Gradients/(counts, sums) are computed in return, buffers are dis-assigned from DMAs and the last ones are destructed.
- **CMF** (Contiguous Memory Free): This call is explicitly used to free all previously allocated buffers.

It is important to note that the above-demonstrated APIs are Spark independent and can be used in any python application.

5.6 Performance Evaluation

As a case study, we built an LR classification model with 784 features and 10 labels and a K-means clustering model with 784 features and 14 centers, using 40 k available training samples, for a handwritten digits recognition problem. The data are provided by Mixed National Institute of Standards and Technology (MNIST) database [17].

To evaluate the performance of the system and to perform a fair comparison, we built a cluster of four nodes based on the Zynq platform and we compared it with four Spark worker nodes using the Intel Xeon cores. Table 5.3 shows the features of each platform.

It is important to note that a single Spark executor JVM process requires most of the available 512 MB RAM on PYNQ-Z1s, placing a restriction on the Spark

Table 5.3 Main features of the evaluated processors

Features	Xeon	Zynq
Vendor	Intel	ARM
Processor	E5-2658	A9
Cores (threads)	12 (24)	2
Architecture	64-bit	32-bit
Instruction set	CISC	RISC
Process	22 nm	28 nm
Clock frequency	2.2 GHz	667 MHz
Level 1 cache	380 kB	32 kB
Level 2 cache	3 MB	512 kB
Level 3 cache	30 MB	–
TDP	105 W	4 W
Operating system	Ubuntu	Ubuntu

application, which requires main memory to cache and repeatedly access the working dataset from FPGAs off-chip RAM once read from HDFS. This results in delays during the execution as inevitably are performed transfers between the memory and the swap file, which is stored inside the SD card. This memory restriction is also the reason why we limit the number of Spark executors to 1 ARM core per node, thus preventing both cores from performing Spark tasks simultaneously.

On the other hand, the Xeon system consists of 12 cores with 2 threads for each core. The Spark cluster started on this platform allocates 4 out of 24 threads, as worker instances, in order to compare it with the 4 nodes of the Pynq cluster.

We also compared the accelerated platform with the software-only scenario in which the algorithm is executed only on the ARM cores. Such comparison is valuable as there are applications, where only embedded processors can be used and big-core systems like Xeon cannot be supported due to power constraints.

5.6.1 Latency and Execution Time

Figure 5.6 depicts the execution time of the Logistic Regression application running on a high-performance ×86_64 Intel processor (Xeon E5 2658) clocked at 2.2 GHz and a Pynq cluster which makes use of the Programmable Logic, for an input dataset of 40,000 lines split in chunks of 5000 lines, for various numbers of GD iterations. The operating frequency of the Zynq FPGAs is 142 MHz, while the on-chip ARM cores are clocked at 666 MHz. As it is shown, the acceleration factor is equivalent to the number of the iterations. An equivalent diagram for K-means application is shown in Fig. 5.7.

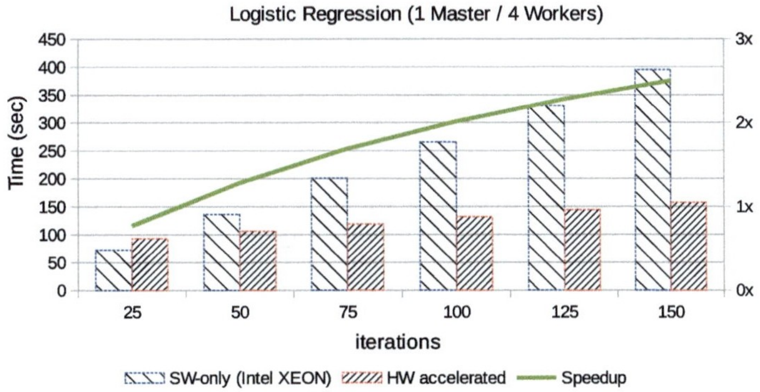

Fig. 5.6 Logistic regression speedup versus the number of the iterations (Intel XEON vs. Pynq)

Table 5.4 Execution time (s) of the worker (mapper) functions

Worker type	Data extraction	GD algorithm computations (per iteration)	K-Means algorithm computations (per iteration)
Intel XEON	7.5	2.6	3.3
ARM	80	46.6	41.5
Pynq	80 (ARM)	0.51 (FPGA)	0.54 (FPGA)

In more detail, in the PYNQ-Z1 boards the data extraction part, for the Logistic Regression application, takes about 80 s to complete, while every iteration of the algorithm is completed in 0.51 s, since the train input data is already cached into the previously allocated buffers. On the other hand, Xeon CPU reads, transforms, and caches the data in only 7.5 s, but every Gradient Descent iteration takes approximately 2.6 s. This is the reason why the speedup actually depends on the number of iterations that are performed. For this specific example the LR model converges, and achieves up to 91.5% accuracy, after 100 iterations of the algorithm, in which up to 2× system speedup is achieved compared to the Xeon processor. However, there are cases in which much higher number of iterations is required, until the convergence criteria is met, and thus much higher speedup can be observed.

For the K-means clustering application, the results are also similar. The data extraction in this application requires the exact same time with the Logistic Regression one, as the same dataset is used, while there is also a significant speedup (up to 2.5×) of the iterative algorithm computations. Again 100 iterations are examined here as the default value that is used in our library (if not specified by the user), however metrics for different values of this parameter are also presented.

Table 5.4 shows the execution time of the main mapper functions, which are executed on the worker nodes. In the Xeon platform and the ARM-only case, both the data extraction and the algorithm computations are performed on the CPUs, while in the Pynq workers the data extraction is executed on the ARM core while the algorithmic part is offloaded to the programmable logic.

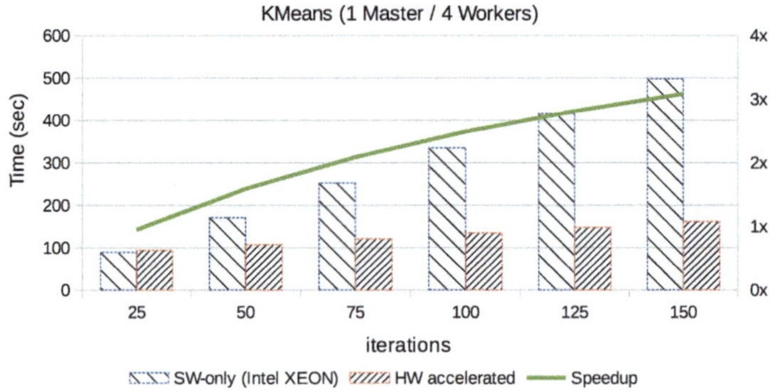

Fig. 5.7 K-means speedup versus the number of the iterations (Intel XEON vs. Pynq)

Fig. 5.8 Logistic regression speedup versus the number of the iterations (ARM vs. Pynq)

Figures 5.8 and 5.9 show the speedup of the accelerated execution compared to the software-only solution running on the same cluster but using only the ARM processors. In this case, we can achieve up to 36× speedup for Logistic Regression and 31× for K-means, compared to the software-only case, which shows that it is definitely crucial to provide accelerator support for future embedded data centers.

5.6.2 Power and Energy Consumption

To evaluate the energy savings, we measured the average power running the algorithm both in the SW-only, and the HW-accelerated cases. In order to measure the power consumption of the Xeon server, we used Intel's Processor Counter Monitor (PCM) API, which, among others, enables capturing the power consumed by the CPU and DRAM memory for executing an application. We also measured the power consumption in the accelerated case using the ZC702 Evaluation board, which hosts

Fig. 5.9 K-means speedup versus the number of the iterations (ARM vs. Pynq)

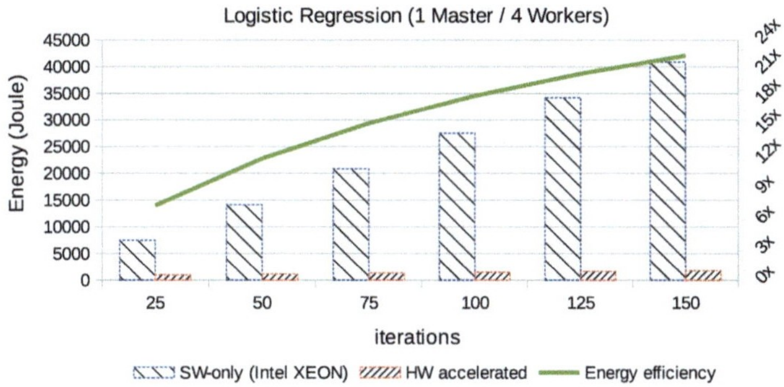

Fig. 5.10 Logistic regression energy consumption based on the number of iterations (Intel XEON vs. Pynq)

the same Zynq device as the PYNQ-Z1 board, taking advantage of the on-board power controllers.

Figures 5.10 and 5.11 show the energy consumption of the Xeon processor compared to the Pynq cluster. The average power consumption of the Xeon processor and the DRAMs is 100 W, while a single Pynq node (both the AP SoC and the DRAM) consumes about 2.6 W during the data extraction and 3.2 W during the hardware computations. In that case, we can achieve up to 23× better energy efficiency due to the lower power consumption and the lower execution time.

In Figs. 5.12 and 5.13 is depicted the energy consumption comparison between the SW-only (ARM) and HW-accelerated execution of the application on the Pynq cluster. It is clear that the average power consumption of the accelerated case is slightly higher than the power consumption of the ARM-only one, because of the need to power supply also the programmable logic. However, due to the significant much higher execution time of the ARM-only solution, eventually, up to 29× lower energy consumption is achieved.

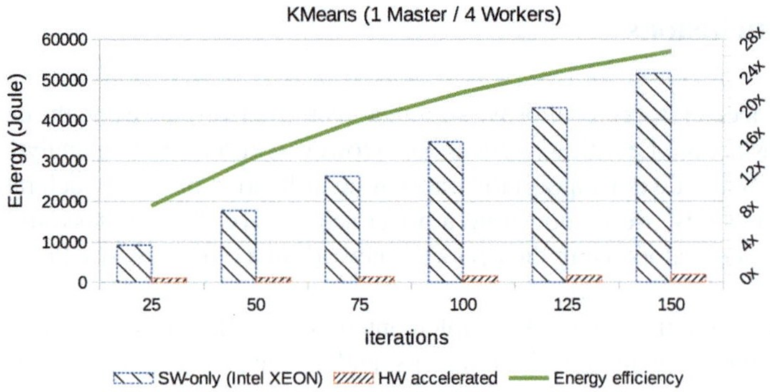

Fig. 5.11 K-means energy consumption based on the number of iterations (Intel XEON vs. Pynq)

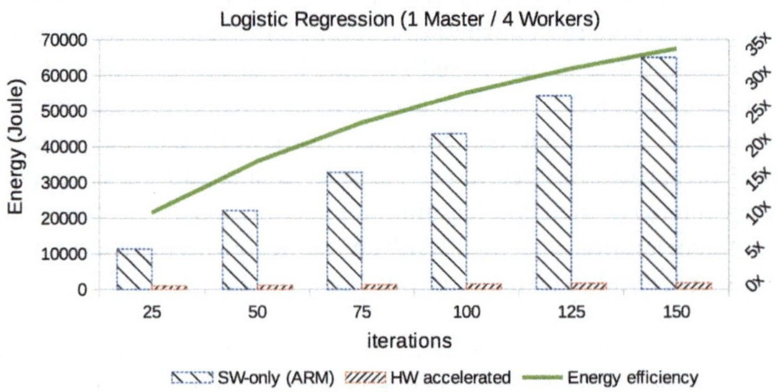

Fig. 5.12 Logistic regression energy consumption based on the number of iterations (ARM vs. Pynq)

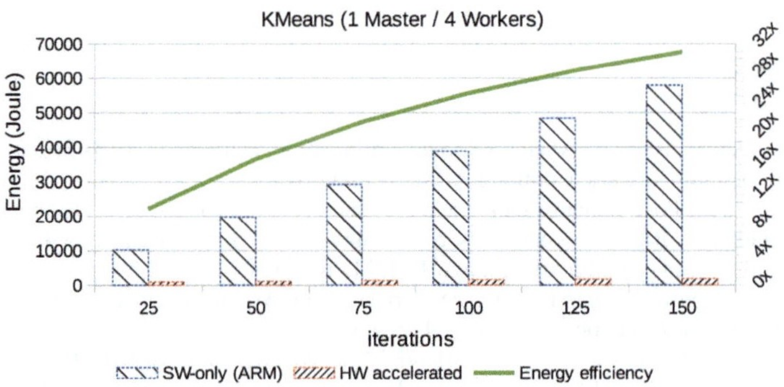

Fig. 5.13 K-means energy consumption based on the number of iterations (ARM vs. Pynq)

5.7 Conclusions

Hardware accelerators can improve significantly the performance and the energy efficiency of Machine Learning applications. However, currently data analytics frameworks like Spark do not support the transparent utilization of such acceleration modules. In this study, we demonstrate a novel scheme for the seamless utilization of hardware accelerators using the Spark framework in a cluster of All-Programmable SoCs.

For the evaluation, we have implemented two hardware accelerators, one for Logistic Regression and one for K-means, and we have efficiently integrated it with Spark. The accelerators can be easily utilized by a Spark user, as the main change that needs to be made in the application, is the replacement of the library that is imported. The results show that the proposed system can be used in high-performance systems to reduce the energy consumption (up to 23×) and also reduce up to 2.5× the execution time, while in embedded systems it can achieve up to 36× speedup compared to the embedded processors and up to 29× lower energy consumption.

The results also show that the proposed framework can be utilized to support any kind of hardware accelerators in order to speedup the execution time of computational-intensive machine learning applications based on Spark, and prove that hardware acceleration and thus SW/HW co-design is in fact a valid solution when software acceleration techniques meet their limits.

Acknowledgements This project has received funding from the European Union Horizon 2020 research and innovation programme under grant agreement No 687628—VINEYARD H2020. We also thank Xilinx University Program for the kind donation of the software tools and hardware platforms.

References

1. Esmaeilzadeh H, Blem E, Amant RS, Sankaralingam K, Burger D (2013) Power challenges may end the multicore era. Commun ACM 56(2):93–102
2. Martin C (2014) Post-dennard scaling and the final years of Moores Law. Technical report
3. Esmaeilzadeh H, Blem E, Amant RS, Sankaralingam K, Burger D, Burger D (2012) Dark silicon and the end of multicore scaling. IEEE Micro 32(3):122–134
4. Ghasemi E, Chow P. Accelerating apache spark with FPGAs. In: Concurrency and computation: practice and experience, pp e4222–n/a. e4222 cpe.4222
5. Huang M, Wu D, Yu CH, Fang Z, Interlandi M, Condie T, Cong J (2016) Programming and runtime support to blaze FPGA accelerator deployment at datacenter scale. In: Proceedings of the seventh ACM symposium on cloud computing, SoCC '16. ACM, New York, NY, USA, pp 456–469
6. Chen YT, Cong J, Fang Z, Lei J, Wei P (2016) When spark meets FPGAs: a case study for next-generation dna sequencing acceleration. In: 2016 IEEE 24th annual international symposium on field-programmable custom computing machines (FCCM), pp 29–29, May 2016
7. Cong J, Huang M, Wu D, Yu CH (2016) Invited—heterogeneous datacenters: options and opportunities. In: Proceedings of the 53rd Annual design automation conference, DAC '16. ACM, New York, NY, USA, pp 16:1–16:6

8. Segal O, Colangelo P, Nasiri N, Qian Z, Margala M (2015) SparkCL: a unified programming framework for accelerators on heterogeneous clusters. arXiv:1505.01120
9. Segal O, Colangelo P, Nasiri N, Qian Z, Margala M (2015) Aparapi-Ucores: A high level programming framework for unconventional cores. In: 2015 IEEE high performance extreme computing conference (HPEC), pp 1–6, Sept 2015
10. Kachris C, Soudris D (2016) A survey on reconfigurable accelerators for cloud computing. In: 2016 26th International conference on field programmable logic and applications (FPL), pp 1–10, Aug 2016
11. Stuecheli J, Blaner B, Johns CR, Siegel MS (2015) CAPI: a coherent accelerator processor interface. IBM J Res Dev 59(1):7:1–7:7
12. Xilinx reconfigurable acceleration stack targets machine learning, data analytics and video streaming. Technical report, 2016
13. Byma S, Steffan JG, Bannazadeh H, Leon-Garcia A, Chow P (2014) FPGAs in the cloud: booting virtualized hardware accelerators with openstack. In: 2014 IEEE 22nd Annual International Symposium on Field-Programmable Custom Computing Machines (FCCM), pp 109–116, May 2014
14. Apache, spark, http://spark.apache.org/
15. Zaharia M, Chowdhury M, Das T, Dave A, Ma J, McCauley M, Franklin MJ, Shenker S, Stoica I (2012) Resilient distributed datasets: a fault-tolerant abstraction for in-memory cluster computing. In: Proceedings of the 9th USENIX conference on networked systems design and implementation, NSDI'12. USENIX Association, Berkeley, CA, USA, pp 2–2
16. Pynq: Pyhton productivity for Zynq. Technical report, 2016
17. LeCun Y, Cortes C (2010) MNIST handwritten digit database

Chapter 6
M2DC—A Novel Heterogeneous Hyperscale Microserver Platform

Ariel Oleksiak, Michal Kierzynka, Wojciech Piatek, Micha vor dem Berge, Wolfgang Christmann, Stefan Krupop, Mario Porrmann, Jens Hagemeyer, René Griessl, Meysam Peykanu, Lennart Tigges, Sven Rosinger, Daniel Schlitt, Christian Pieper, Udo Janssen, Holm Rauchfuss, Giovanni Agosta, Alessandro Barenghi, Carlo Brandolese, William Fornaciari, Gerardo Pelosi, Joao Pita Costa, Mariano Cecowski, Robert Plestenjak, Justin Cinkelj, Loïc Cudennec, Thierry Goubier, Jean-Marc Philippe, Chris Adeniyi-Jones, Javier Setoain and Luca Ceva

6.1 Introduction

In recent years, the market of server platforms and solutions for data centers is quickly evolving to follow the fast pace of data center capacities' growth. This growth is needed to serve demands for cloud services and huge network traffic. These needs are mostly imposed by emerging applications and technologies such as advanced mobile devices, Internet of Things (IoT), 5G, virtual and augmented reality (VR/AR), machine learning, and artificial intelligence.

To cope with challenges and scale needed for such demanding applications, data center operators and technology vendors provide new solutions and advance best

---

A. Oleksiak · M. Kierzynka · W. Piatek
Poznan Supercomputing and Networking Center, ul. Noskowskiego 10, 61-704 Poznan, Poland
e-mail: ariel@man.poznan.pl

M. Kierzynka
e-mail: michal.kierzynka@man.poznan.pl

W. Piatek
e-mail: piatek@man.poznan.pl

M. vor dem Berge · W. Christmann · S. Krupop
Christmann Informationstechnik + Medien GmbH & Co. KG, Ilseder Huette 10c, 31241 Ilsede, Germany
e-mail: micha.vordemberge@christmann.info

W. Christmann
e-mail: wolfgang.christmann@christmann.info

S. Krupop
e-mail: stefan.krupop@christmann.info

C. Kachris et al. (eds.), *Hardware Accelerators in Data Centers*,
https://doi.org/10.1007/978-3-319-92792-3_6

practices to ensure high availability, efficiency, and performance. This leads to creation of hyperscale data centers. Their scale allows to use very efficient technologies, often dedicated solutions and hardware, and make it economically viable. On the other hand, recently the paradigm of the so-called edge data centers (putting smaller data centers closer to local data) gains importance as it is an answer to huge amounts of data being generated and requested by mobile end users and emerging IoT systems. Although hyperscale and edge solution are opposite approaches, they have common challenges to address. Both hyperscale and edge solutions require high efficiency, low environmental impact as well as advanced monitoring with self-healing and self-optimization functions to lower costs of maintenance and ensure high availability.

M. Porrmann (✉) · J. Hagemeyer · R. Griessl · M. Peykanu · L. Tigges
Cognitronics and Sensor Systems Group, CITEC, Bielefeld University, Bielefeld, Germany
e-mail: mporrmann@cit-ec.uni-bielefeld.de

J. Hagemeyer
e-mail: jhagemeyer@cit-ec.uni-bielefeld.de

R. Griessl
e-mail: rgriessl@cit-ec.uni-bielefeld.de

M. Peykanu
e-mail: mpeykanu@cit-ec.uni-bielefeld.de

L. Tigges
e-mail: ltigges@cit-ec.uni-bielefeld.de

S. Rosinger · D. Schlitt · C. Pieper
OFFIS e. V. – Institute for Information Technology, Oldenburg, Germany
e-mail: sven.rosinger@offis.de

D. Schlitt
e-mail: Daniel.Schlitt@offis.de

C. Pieper
e-mail: christian.pieper@offis.de

U. Janssen
CEWE Stiftung & Co. KGaA, Oldenburg, Germany
e-mail: udo.janssen@cewe.de

H. Rauchfuss
Huawei Technologies, German Research Center, Riesstrasse 25, 80992 Munich, Germany
e-mail: holm.rauchfuss@huawei.com

G. Agosta · A. Barenghi · C. Brandolese · W. Fornaciari · G. Pelosi
DEIB – Politecnico di Milano, Piazza Leonardo da Vinci 32, Milano, Italy
e-mail: giovanni.agosta@polimi.it

A. Barenghi
e-mail: alessandro.barenghi@polimi.it

C. Brandolese
e-mail: carlo.brandolese@polimi.it

W. Fornaciari
e-mail: william.fornaciari@polimi.it

This is caused by the fact that both in case of huge infrastructure and when using many smaller remote data centers the maintenance is difficult and power usage plays an important role.

Another trend is that the required performance and efficiency can be delivered by customization of heterogeneous systems to the needs of specific yet important classes of applications. This customization must be, however, as easy and cheap as possible to make it possible to apply in a large scale.

M2DC aims to address these needs by providing a flexible microserver platform that can be applied to develop various types of appliances optimized for specific classes of applications. As specified in [14], the targeted features of M2DC appliances include rich possibility of customization, energy efficiency, relatively low cost, and facilitated maintenance (cf. Fig. 6.1).

Main advantages of the appliances that are developed in M2DC include a flexible hardware platform based on standard interfaces and an innovative design that allows integration of a wide spectrum of microservers: ARM-based multi/many-cores, typical x86 modules, Multiprocessor System-On-Chips (MPSOCs), reconfigurable devices (FPGAs), and GPUs. Additionally, the architecture introduces a

G. Pelosi
e-mail: gerardo.pelosi@polimi.it

J. Pita Costa · M. Cecowski · R. Plestenjak · J. Cinkelj
XLAB d.o.o., Pot za Brdom 100, 1000 Ljubljana, Slovenia
e-mail: joao.pitacosta@xlab.si

M. Cecowski
e-mail: mariano.cecowski@xlab.si

R. Plestenjak
e-mail: robert.plestenjak@xlab.si

J. Cinkelj
e-mail: justin.cinkelj@xlab.si

L. Cudennec · T. Goubier · J.-M. Philippe
CEA, LIST, PC 172, 91191 Gif-sur-Yvette CEDEX, France
e-mail: loic.cudennec@cea.fr

T. Goubier
e-mail: thierry.goubier@cea.fr

J.-M. Philippe
e-mail: jean-marc.philippe@cea.fr

C. Adeniyi-Jones · J. Setoain
ARM Ltd., CPC-1 Capital Park, Fulbourn, Cambridge CB21 5XE, UK
e-mail: chris.adeniyi-jones@arm.com

J. Setoain
e-mail: javier.setoain@arm.com

L. Ceva
Vodafone Telematics, Varese, Italy
e-mail: luca.ceva@vodafonetelematics.com

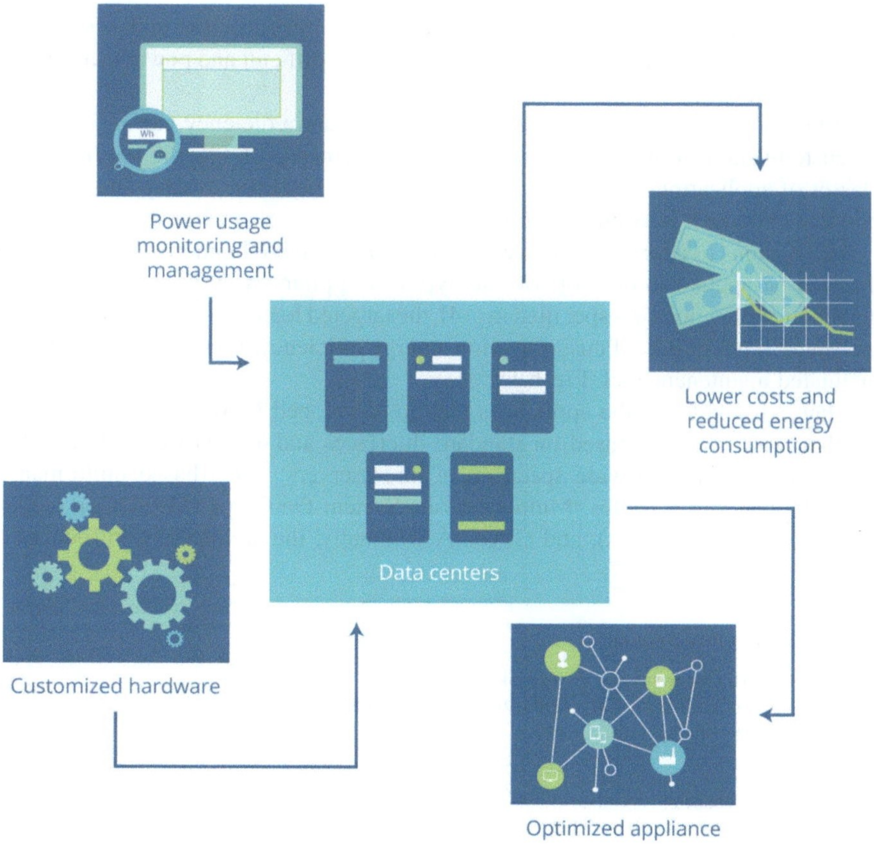

Fig. 6.1 With appliances based on its heterogeneous microserver platform, M2DC targets optimization of energy efficiency, performance, and TCO

concept of System Efficiency Enhancements (SEEs)—built-in functions that enable the use of hardware acceleration in a seamless way. The main role of SEEs is to enhance efficiency, performance, reliability, or security of the micro-server platform at possibly lowest overhead in terms of performance, power, and configuration effort. Examples of SEEs developed within M2DC include advanced platform monitoring, pattern matching for events in the system, communication encryption, image processing, artificial neural networks, and low-latency communication.

On top of the hardware and firmware layer, M2DC develops a middleware and software tools helping to efficiently manage the platform. In this way, the M2DC appliances go beyond the hardware into the whole software stack and ecosystem of tools. To enable easy integration into existing data centers, the middleware contains extensions of and components with interfaces to OpenStack [15]. Additionally, special extensions and/or integration work has been done on other popular tools, for instance, for monitoring Zabbix, queuing systems SLURM, etc. M2DC develops a suite of tools and algorithms for node provisioning and intelligent power and thermal management.

To verify the flexibility and benefits of the platform, several appliances for important usage scenarios are being developed, namely for specific classes of applications such as image processing and IoT data analytics as well as for Platform as a Service (PaaS) clouds and High Performance Computing (HPC) centers.

Organization of the chapter The rest of this chapter is organized as follows. In Sect. 6.2, we provide an overview of the M2DC modular microserver hardware and software architecture. In Sect. 6.3, we describe the range of application scenarios, which are used to verify and demonstrate the M2DC technology. Finally, in Sect. 6.4, related work in other European projects is reviewed, focusing on projects that build the baseline for the developments within M2DC, and in Sect. 6.5 we draw our conclusions.

6.2 Architecture Overview

In addition to a steady increase in performance and energy efficiency, customers and applications for HPC and cloud appliances request heterogeneous platforms that can be tailored to their specific requirements, thus enabling further optimization of system efficiency and performance. Therefore, important features of the heterogeneous hyperscale server platform that is developed within the M2DC project include high resource efficiency combined with high scalability, high density and modularity, enabling easy adaptation of the system toward the workload requirements of a wide variety of applications. Compute modules range from low-power architectures to high-performance microservers, which can be extended by reconfigurable and massively parallel hardware accelerators utilizing a dedicated high-speed, low-latency communication infrastructure. Within the project, efficiency in terms of performance, energy, and TCO will be demonstrated by a representative mix of turn-key appliances that are supported by an intelligent, self-optimizing management infrastructure.

Based on the customer requirements and on the facilities of the data centers, the M2DC platform can be easily configured, providing low-power solutions that can be easily integrated into existing data center environments on the one hand and providing unparalleled density of server nodes in new data center environments, offering the required cooling and power facilities, on the other hand. Although developed in an EU project, the M2DC modular microserver system architecture is not just a research platform, but also targets reliability and maintainability levels of a commercial product. A blade-style system approach is chosen, easing maintenance and providing hot-swap and hot-plug capabilities (cf. Fig. 6.2).

Communication facilities will be provided on three levels within the M2DC server: On level one, a dedicated monitoring and control network provides facilities for management of the core functionalities based on a reliable embedded bus connecting all parts of the M2DC server system as well as an integrated IPMI/I-KVM solution for every microserver. On level two, a full-featured Ethernet-based communication

Fig. 6.2 The M2DC server combines up to 45 high-performance or 240 low-power microservers in a single 3 RU chassis

backbone providing multiple 1 and 10 Gbit/s links to each microserver is integrated. A key feature of M2DC is the third level of communication, integrating a dedicated infrastructure for high-speed, low-latency communication into the M2DC server. Based on high-speed serial transceivers and PCIe switching, this infrastructure provides high-speed, low-latency communication between the microservers.

The monitoring and control system manages all core functions of the M2DC server system, i.e., it controls the power states of all components, enables a controlled system startup and shutdown, and manages the multilevel communication infrastructure according to the need of applications or appliances. In addition, it monitors the health of the server unit by providing fast and easy access to the more than 10,000 sensor values in a single rack (e.g., power, voltage, temperature on device level, microserver level, as well as server level). An integrated distributed network of microcontrollers is used for preprocessing the huge amount of sensor data, providing the user exactly the information he requires and enabling smart and effective power and energy management solutions. Therefore, new deployment and management technologies are combined with a proactive power management for providing QoS-aware dynamic performance settings, exploiting the heterogeneity of the server platform. Additionally, new methods for thermal management are developed focusing on the specific requirements and capabilities of heterogeneous architectures, enabling runtime adaptation of their behavior at all levels, ranging from a single microserver to the complete data center. Based on the actual requirements, optimization goals can vary from performance maximization to minimization of power/energy requirements or hotspot avoidance.

ARMv8-based multicore processors both from the server domain and from the mobile domain significantly increase the energy efficiency compared to state-of-the-art x86-based server platforms. Therefore, they have been selected as one of the key components supported by the M2DC server platform, offering a good compromise between energy efficiency and ease of programming. When targeting reduces in energy consumption by an order of magnitude or even more, CPU-based approaches typically are outperformed by massively parallel or reconfigurable architectures like

GPUs, MPSoCs, or FPGAs. It has been shown already that these architectures can provide the required performance/power ratios but typically they are integrated as hardware accelerators, directly attached to a dedicated host and optimized for a specific application. The M2DC server targets a more flexible approach, integrating dynamically reconfigurable FPGAs as well as GPUs for System Efficiency Enhancements (SEE) targeting both application acceleration and system level functionalities like efficient integrated security features and increased dependability. On hardware level, the flexible integration of accelerators via the high-speed, low-latency network enables pooling of accelerator resources as well as assignment to changing hosts, even at runtime.

For proving the efficiency of the platform and its supporting software infrastructure, turnkey appliances are developed for selected applications providing an optimized mapping between software and heterogeneous hardware components. Depending on the application requirements, the corresponding appliances are based on different middleware layers and techniques, mostly running directly on the operating system without virtualization overhead. The targeted "bare metal cloud" approach dynamically installs a desired operating system combined with the required libraries and applications onto physical nodes. Nevertheless, if an appliance benefits more from the flexibility of a virtualized and containerized cloud-like approach, this is also supported within M2DC. A base appliance providing an Infrastructure-as-a-Service (IaaS) and Metal-as-a-Service (MaaS) layer is provided as the basis for the personalized appliances and can be directly provided to users who want to install their own individual applications.

The targeted applications are analyzed in M2DC range from general cloud computing via image processing and big data analytics to HPC applications, representing a wide variety of different requirements, as discussed in Sect. 6.3. To improve usability, reuse, and user acceptance, all relevant interfaces like the operating system management and the appliance management are based on widely used standards. This includes required interfaces for smooth integration with DCIM and HPC management software allowing fine-grained monitoring and a comprehensive set of power management functions.

6.2.1 M2DC Server Architecture

The M2DC architecture enables the integration of different types of microservers. While existing microserver platforms mostly support a single, homogeneous microserver architecture, M2DC supports the full range of heterogeneous microserver technology from CPUs to FPGAs, which can be seamlessly combined into a single chassis. The M2DC server architecture supports microservers based on x86 (e.g., Intel Xeon), 64-bit ARM mobile/embedded SoCs, 64-bit ARM server processors, FPGAs, and GPUs, as well as other PCIe-based acceleration units. This heterogeneity can be used to configure and build an optimized processing platform based on the needs of the application. As shown in Table 6.1, all the main microservers (CPU,

Table 6.1 Selected M2DC microservers

	CPU	GPU	FPGA
Low-power microserver (Jetson/Apalis)	NVIDIA Tegra X1 2 Core Denver + 4 Core A57 + Pascal GPGPU@1.5 GHz		Xilinx Zynq 7020 85 kLC
High-performance microserver (COM Express)	ARMv8 Server SoC 32 Core A72@2.4 GHz		
	Intel Xeon E3 4 Core Kaby Lake@3.0 GHz	NVIDIA Tesla P100 Pascal GPGPU@1.3 GHz	Intel Stratix 10 SoC 1,092 kLC
	Intel Xeon D 16 Core Broadwell@1.3 GHz		

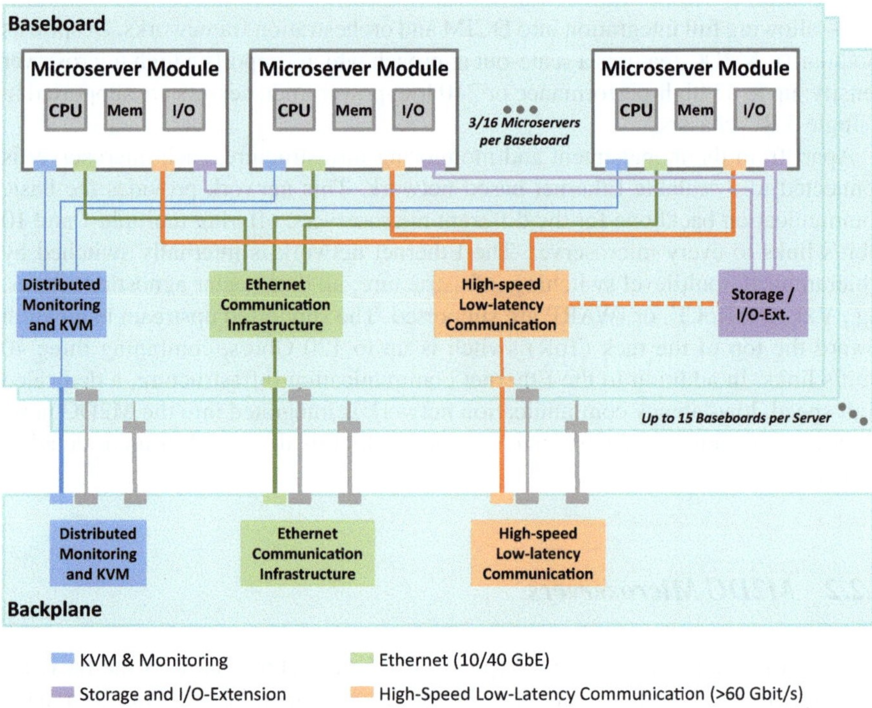

Fig. 6.3 Overview of the M2DC server architecture

GPU, and FPGA) are available in a low-power as well as in a high-performance variant. Like the big-little approach in today's mobile processors, this feature allows to increase the energy efficiency even more by dynamically switching, e.g., between 64-bit ARM server processors and 64-bit ARM mobile SoCs depending on the current load situation of the application.

As depicted in Fig. 6.3, the M2DC server features a modular approach. This modularity ensures flexibility and reusability, thereby ensuring high levels of maintainability. Microservers are grouped on a baseboard or carrier blade, which supports hot-swapping and hot-plugging of microservers, similar to a blade-style server. Using existing computer-on-module or server-on-module form factors, which are established in the industry, allows reuse of already existing microserver developments. This eases integration of third-party microserver modules, providing a broad set of commercial of the shelf microserver modules readily available for usage in the M2DC server. Fine-grained power monitoring and control within the system on hardware level enables sophisticated high-level management of power, performance, and temperature. The power supplies of the M2DC system are foreseen to be shareable on rack level. Exploiting this feature enables improved energy efficiency as well as higher system reliability due to added redundancy. The distributed monitoring, control, and maintenance infrastructure are accessible via a web interface and a RESTful

API—allowing full integration into DCIM and orchestration frameworks. Despite its modularity, M2DC features a scale-out approach which supports a high microserver density: up to 45 high-performance or 240 low-power microservers are supported in a single 3 RU chassis.

Apart from the management and monitoring infrastructure, each microserver is connected to a scalable Ethernet-based network. This network provides the basic communication backbone for the different microservers, offering multiple 1 and 10 Gbit/s links to every microserver. The Ethernet network is internally switched by a hierarchical, multilevel switching infrastructure; all data center agnostic features, e.g., VxLAN, RoCE, or iWARP are supported. The supported upstream bandwidth toward the top of the rack (ToR) switch is up to 120 Gbit/s, combining three 40 Gbit/s links. In addition to the Ethernet communication infrastructure, a dedicated high-speed, low-latency communication network is integrated into the M2DC next-generation modular microserver architecture, which is described in more detail in Sect. 6.2.3.

6.2.2 M2DC Microservers

As shown in Table 6.1, the COM express [16] form factor is used as the basis for all high-performance microservers, while the Jetson standard from NVIDIA and the Apalis standard from Toradex [19] are used for the low-power microservers. The tables list some examples of possible modules, and many others are available. The COM express form factor supports compact modules of just 125 mm × 95 mm. COM express type 6 and type 7 modules are supported, enabling direct integration of commercial of the self-modules, e.g., x86 modules based on Intel's Skylake/Kabylake architecture. Two new high-performance microservers are developed within M2DC, targeting highly resource-efficient platforms for next-generation data centers, utilizing FPGAs and ARM-based server processors, respectively.

The high-performance ARMv8 microserver is based on an ARM 64-bit SoC, integrating 32 Cortex-A72 cores running at up to 2.4 GHz. The memory controller supports four memory channels populated with DDR4 SO-DIMMs running at 2133 MHz. In total, each microserver integrates up to 128 GB RAM. For connectivity, the microserver provides two 10GbE ports with RoCE support, in addition to a GbE port which is mainly used for management purposes. Up to 24 high-speed serial lanes are available for connection of peripherals or for high-speed, low-latency communication to other microservers in the M2DC server. Using these high-speed links, also multi-socket configurations of the ARMv8 microserver are supported. Additionally, a wide variety of fixed-function units are integrated into the SoC, providing highly resource-efficient acceleration of compression/decompression or security algorithms like asymmetric encryption.

FPGAs are becoming more and more attractive in HPC and cloud computing due to their potentially very high-performance combined with moderate power requirements. High-level synthesis and OpenCL support, which is becoming more and more mature, opens additional application scenarios since programming is no longer limited to hardware specialists. The FPGA-based high-performance microserver that is developed within M2DC will be a full-featured COM Express module, comprising an Altera Stratix 10 SoC with an integrated 64-bit quad-core ARM Cortex-A53 processor. Dedicated DDR4 memory is provided for the CPU as well as for the FPGA fabric, supporting up to four memory channels and up to 64 GB. The high-speed transceivers integrated in the FPGAs are used for PCIe interfacing to communicate with other processor modules, and for high-speed, low-latency communication between high-performance FPGA-based microserver modules.

SoCs targeting the mobile market are promising platforms for data centers when focusing on energy efficiency, especially due to a large amount of integrated accelerators, including GPGPUs, fixed function units, e.g., for video transcoding or even FPGAs. The M2DC server enables integration of modules based on the Jetson standard from NVIDIA, which is used for the currently available Tegra SoCs (Tegra-X1) and the upcoming generations from NVIDIA. Additionally, the Apalis standard from Toradex [19] is supported. With its small form factor of just 82 × 45 mm, it allows a very high density of microservers. In addition to commercially available Apalis modules from Toradex, modules have been developed at Bielefeld University integrating Samsung Exynos5250 SoCs and Xilinx Zynq7020, respectively [11].

6.2.3 High-Speed Communication

In today's implementations, a hardware accelerator is typically attached physically to the PCIe lanes of a CPU node. Using the high-speed, low-latency communication infrastructure of the M2DC server, a hardware accelerator can be flexibly attached to any node within the system. In contrast to state-of-the-art implementations, the communication infrastructure can be used not only to connect CPUs to hardware accelerators but also CPUs to CPUs or accelerators to other accelerators. Furthermore, it is possible to divide the links of certain accelerators, e.g., connecting an accelerator to both a CPU and another accelerator. Thus, accelerators can be combined into a large virtual unit. At runtime, the communication topology can be reconfigured and adapted to changing application requirements via the middleware. Apart from point-to-point communication, the flexible communication infrastructure also allows efficient multi- or broadcast communication topologies.

In more detail, the high-speed, low-latency communication infrastructure of the M2DC server is based on two technologies: asynchronous crosspoint switches that allow connections between microservers independent of the used protocol and PCI Express switches, which are used for packet routing based on PCI Express. These two technologies are combined on baseboard level as well as on the chassis backplane. Additionally, the whole communication infrastructure can be scaled across rack level by using additional connectors located at the back panel. In addition to direct communication between the different microservers via PCIe, the platform also supports connection to storage or I/O extensions. This enables easy integration of

PCIe-based extension cards like GPGPUs or storage subsystems, which can also be shared across multiple microservers via the multi-root I/O virtualization (MR-IOV) feature of the PCIe communication infrastructure.

As mentioned above, in addition to PCIe, the M2DC server supports direct links between microservers, using asynchronous crosspoint switches independent of the used protocol. This feature is of particular interest for FPGAs, as these devices support low-level point-to-point communication schemes not involving the overhead of protocols like PCIe. For example, use cases include a direct communication infrastructure between multiple FPGAs as well as multiple PCIe endpoints implemented on the FPGA, used for communication toward the processor-based microservers.

6.2.4 System Efficiency Enhancements

System Efficiency Enhancements (SEEs) are integrated into the M2DC server platform to increase its efficiency, e.g., with respect to performance or energy but also targeting reliability or security. The SEEs are seamlessly integrated hardware/software components that utilize the embedded FPGA and GPU accelerators together with the flexible high-speed interconnect to provide a wide variety of mechanisms for global system efficiency enhancements. Three categories of SEEs are developed in the project: communication SEEs dedicated to acceleration of communication and runtime functions, application SEEs dedicated to accelerating application-domain compute kernels, and system SEEs dedicated to enhancing the platform efficiency and reliability. Depending on the actual requirements, the accelerators can dynamically adapt their behavior, e.g., toward performance improvements, power reduction, and dependability.

Communication SEEs target optimization of the internal communication within the M2DC server, e.g., data transfer and synchronization between tasks or processes of an application. Envisioned SEEs include MPI acceleration, remote DMA, cluster-wide available low-latency global scratchpad memory as well as acceleration of various cache coherency protocols. From the application level, these SEEs are fully transparent.

In M2DC, application SEEs are typically implemented using OpenCL, CUDA, or RTL/HLS. Based on the targeted appliances, first implementations include FPGA-based accelerators for image processing. Additionally, machine learning is targeted, on the one hand, based on deep learning and on the other hand based on self-organizing feature maps, especially utilizing the high-speed, low-latency communication of the M2DC server platform to enable efficient scalability for large neural network implementations.

System SEEs include enhancements for monitoring, security, and reliability that are provided independent of the actual application. These SEEs will, e.g., assist collecting and (pre-)processing of monitoring data in real time allowing runtime optimizations based on in-system data mining since optimal operation of an appliance may depend on specific ambient characteristics (temperature, power, hot-spots, etc.). Security is enabled by encryption SEEs, both symmetric and homomorphic encryptions, as well as an intrusion detection SEE. Efficient OpenCL implementations of cryptographic primitives [1–3], which can be transparently deployed on GPUs or on FPGAs, are developed utilizing OpenCL-based synthesis tools. This strategy allows optimization of a single efficient implementation, which is transparently deployed on any targeted hardware provided as a service, allowing also higher flexibility, should need to change the employed cryptographic primitives arise.

6.2.5 Middleware Stack

Building on the Linux operating system (e.g., Linaro for ARM-based compute modules) and other well-known software infrastructures, M2DC will also feature optimized runtime software implementations when needed, to improve the efficiency of the system toward application domains such as cloud computing, big data analytics, and HPC applications. At the heart of the middleware for "bare metal cloud" sits OpenStack Ironic [15], which provides bare metal (micro)server software deployment and lifecycle management for those. OpenStack Ironic will be modified and complemented by other OpenStack components for handling the dynamic and heterogeneous nature of the M2DC microserver nodes, in particular, the hardware accelerators. It will interact with the M2DC chassis controller via a DMTF Redfish API.

Due to its heterogeneity, power density, and thus, possible thermal imbalance, the M2DC microserver requires advanced resource and thermal policies that allow achieving significant energy savings while maintaining high reliability under various conditions. These conditions and challenges include workload fluctuations, managing hot spots, power leakage, and finding a trade-off between workload and resource management. To address these issues, energy- and thermal-aware techniques benefit from a set of sensors located inside the server and monitoring tools together with corresponding power and thermal models used to determine trends in their changes. These models are supported with benchmarking data and statistical method increasing their accuracy. By analyzing collected data and predicting future trends, the resource and thermal management module performs energy optimizations. The optimization methods consist of dynamic power management of M2DC microserver components including actions like switching particular modules off, management of processors' C-states, and dynamic addition/removal of cores. Moreover, dynamic voltage and frequency scaling (DVFS) is applied (if possible) for adjusting the speed of individual processors to meet thermal constraints. The resource and thermal management module also supports power capping mechanisms that allow users to provide limitations for the maximum power drawn by the system. Finally, a dedicated fan controller is in charge of adjusting the fans speed in order to keep all the components within the desired temperature range and optimize their power usage. This is done continuously, in a proactive way and for each fan separately. All these components take their control actions through a dedicated monitoring system and OpenStack services responsible for reading sensor values and providing all information to the management modules.

6.3 Use Case Scenarios

One of the most important M2DC objectives is to deliver appliances well suited for specific classes of popular or emerging relevant applications. Thus, a crucial part of the strategy of the project is to steer the joint development of both software and hardware by specific real-life use cases identified by the project partners. Such a continuous validation will enable M2DC to deliver optimized appliances customized to relevant and real-life workloads and use cases. The M2DC use cases have been carefully selected to ensure their potential-wide uptake, market relevance, and special importance for future computing and data processing needs. In the following, the use cases targeted in M2DC are briefly discussed, describing their goals, technical aspects, and how they benefit from the project.

6.3.1 Low-Cost Image Processing

Low-cost image processing is a use case based on actual requirements from the Online Photo Services (OPS) of CEWE, where customers are able to prepare and order photo print products online, using a web browser. All requested image processing operations like scaling, cropping, or rotating are performed in data centers, independent of the customer's hardware. To satisfy the commonly high usability standards of on-demand services, all tasks must meet strict response time constraints, requiring a fast execution of image processing tasks and therefore constantly available computing resources. Due to high workload variance based on fluctuating user demand, there are times with low workloads offering great energy saving potentials. However, as workload varies even on a daily basis (e.g., peak in evening vs. bottom at night), a static workload redistribution approach is not sufficient.

For this purpose, the heterogeneous hardware approach of M2DC will be utilized to realize a flexible and more energy-efficient image processing appliance that benefits from different power/performance ratios. The image processing tasks will be adopted to run on x86, ARM, and FPGA/GPGPU hardware, controlled by a service-oriented API. An intelligent workload management deploys queued tasks on the most efficient hardware and even controls server states using long-term forecasts based on historical workload data, to reduce the amount of idling servers.

While the mentioned approach is rather specific, the low-cost image processing will also be offered as a more generic appliance for efficient on the fly thumbnail generation for a large amount of images. This can be integrated into existing data center ecosystems more easily, addressing a wider range of customers.

6.3.2 IoT Data Analytics for Transport

The transport application domain is characterized by a vast amount of data that are currently underused—each vehicle (cars, motorbikes, and trucks) can be sensorized to collect a wide range of data, such as GPS position and accelerometer readings. Currently, non-automotive companies are offering a support service that automatically detects accidents and sends requests for assistance, by leveraging an IoT system composed of a sensing unit that includes an accelerometer (to detect crashes), a GPS unit providing localization, and a data processing and communication unit. The same companies are expanding the service to collect a wider range of data beyond accident detection, with the goal of providing a full driver behavior profiling, which can be used to offer added-value services, such as personalized insurance pricing. Such extension implies a massive increase in the amount of stored and processed data. To this end, the M2DC project aims at offering a solution for optimizing the TCO of cloud servers employed for the storage and processing of IoT data, while maximizing the amount of data processed within a given power/energy budget [14].

The strategy adopted in M2DC is to leverage SEEs for data analytics to optimize the execution of driver behavior profiling. These analyses process the data from a set of cars, extracting features such as aggressive driving (characterized by exceeding acceleration thresholds) or overspeeding behaviors (the propension to ignore speed limits), to classify trips performed with a single car as belonging to different drivers, or to classify drivers in different categories based on the driver behavior. As an example of such an application, consider the identification of different drivers sharing the same car. It is possible to extract from the collected data information about the behavior of the driver, such as brusque acceleration or deceleration patterns, propension to speeding, comparing them to the average behavior of that car on a given trip (taking into account the type of road and the time of day to distinguish behavioral changes induced by, e.g., traffic or road conditions). By applying classification algorithms such as k-means on such features, it is possible to identify whether the behaviors belong to a single driver or to multiple drivers.

The underlying technology is a modular compilation and runtime toolchain (derived from LLVM) interfacing the R language to an heterogeneous data flow execution runtime (currently based on StarPU runtime software [5]). A dedicated version of the compilation toolchain will be developed on top of M2DC hardware and middleware, including new transformation and analysis passes at compiler level. The existing runtime will be extended to support the hardware and to use capabilities of the hardware/middleware pair. The underlying runtime, similarly to the existing one, will also be used to build programs without any compiler support and the transformation passes will be performed at IR (Intermediate Representation) level, enabling the use of other languages to target the same toolchain at a lower cost (only the front end would need to be rewritten).

6.3.3 Energy-Optimized Platform as a Service

Platform as a Service (PaaS) allows customers to run their applications on a full-stack environment, and therefore the service needs to support the most popular programming languages and web applications. In this scenario, the computing power is delivered by isolated runtime containers that are spawned on demand by cloud orchestration platforms. Low-power servers and precise power and thermal management are crucial for keeping low-energy costs while ensuring high availability and required QoS. The cloud appliance is designed to meet these requirements.

6.3.4 High-Performance Computing

High-performance computing (HPC) software tools are among the most time, energy, and resource consuming types of applications. Therefore, it is vital to address this domain, especially as M2DC is all about reducing energy consumption and total cost of ownership. A good representative in this area is EULAG—a numerical HPC solver offering a large spectrum of application fields, such as orographic flows, urban flows, simulation of contamination dispersion, investigation of gravity waves, and many others [7]. EULAG is currently used at Poznan Supercomputing and Networking Center for different scenarios, e.g., air quality monitoring and precise weather prediction. As mentioned above, EULAG is a good example of HPC application as it implements stencil-based computations, requires a low-latency network, and can be efficiently parallelized on modern computational architectures [8, 18]. Therefore, along with traditional HPC benchmarks, it will help to evaluate the computational and networking aspects of the M2DC platform. Based on this, TCO as well as energy efficiency optimized hardware configurations will be proposed for the HPC appliance. Importantly, the appliance will be easy to integrate with a typical data center ecosystem (dimensioning, cooling, networking, and software interfaces) and will be shipped with optimized software libraries for HPC, e.g., MPI with RDMA support, BLAS, etc. It is expected that due to optimized reconfigurable communication and the use of hardware acceleration, the HPC appliance will greatly improve the performance per Watt for complex distributed applications.

6.3.5 Machine Learning

As a high-impact benchmarking application with a high level of parallelism, neural network models are foreseen to be ported and evaluated on the M2DC platform. Even though the project does not officially work on a dedicated machine learning appliance, application SEEs for important neural network algorithms are developed and integrated into the M2DC server to prove the efficiency of it also in this vital area.

Two neural network implementations are developed within the project. On the one hand, an FPGA-based accelerator for deep neural networks is integrated into the platform, capable of efficiently accelerating a wide variety of neural network computing chains including various classifiers like radial basis functions with Gaussian activation function, and multilayer perceptrons with sigmoid, tanh or rectified linear functions. On the other hand, variants of self-organizing feature maps, an efficient tool for data mining [12], will be implemented and evaluated. This implementation is especially targeting large neural networks, thus utilizing the concept of pooling FPGA resources of multiple microservers in one large virtual accelerator.

Additionally, the emulation of spiking neural networks, as developed, e.g., in the Human Brain Project [4], is targeted as a further benchmark for the M2DC server platform. Like HPC EULAG, the neural network implementations will be used to evaluate the scalability of architecture, specifically focusing on the impact of the integrated low-latency communication.

6.4 Related Works

To achieve its ambitious objectives, M2DC takes as baseline the results of different EU projects, in which some of the members of M2DC participate.

Based on the results of FiPS [11, 13], the management of heterogeneity in the server infrastructure is developed. Additionally, M2DC uses lessons learned from the RECS3.0 prototype that has been developed within FiPS as a basis for the development of the M2DC server. RECS3.0 in turn is based on RECS2.0, a result of CoolEmAll [6]. Since RECS3.0 is also based on the computer on module standards COM Express and Apalis, the platform is used for early evaluation of new microserver module developments as well as for software development in the M2DC testbeds

The Mont-Blanc project [17] used commodity energy-efficient embedded technology for building a prototype high-performance computing system based on ARMv7 (32-bit) System-on-Chips. Next-generation system architecture designs based on ARMv8 (64-bit) technology are explored in the Mont-Blanc 2 and Mont-Blanc 3 projects. These projects have been pivotal in building up the software ecosystem required for ARM-based HPC: system software, networking and communication libraries, and support for heterogeneous processing and compute acceleration, much of which is also required for server applications.

The FP7 project EUROSERVER [9] advocates the use of state-of-the-art low-power ARM processors in a new server system architecture that uses 3D integration to scale with both the numbers of cores, and the memory and I/O. While M2DC does not design new chips, the M2DC server is an ideal platform for easy integration of new processor and SoC developments. In addition to standardized interfaces and form factors, the flexible high-speed, low-latency communication infrastructure enables tight yet flexible coupling of new compute resources.

Results from the FP7 project DEEP-ER [10] concerning the benefits of nonvolatile memory storage are of high interest for high-end versions of M2DC appliances, especially those aiming at executing HPC applications. Additionally, advanced checkpointing techniques could improve the reliability of M2DC appliances, if applicable to microserver-based systems.

6.5 Conclusions

The EU H2020 project M2DC aims at the definition of a modular microserver architecture for future data centers targeting a TCO reduction of 50% compared to other servers for selected applications and an improvement in energy efficiency by 10–100 times, compared to 2013 typical servers. As discussed in this chapter, the new heterogeneous M2DC server combines resource-efficient microserver modules with a multilayer communication infrastructure in a highly flexible manner, enabling seamless integration of state-of-the-art x86 processors, 64-bit ARM mobile/embedded SoCs, 64-bit ARM server processors, FPGAs, and GPUs. New microservers are developed within the project integrating, e.g., a 32-core ARMv8 server processor and an Intel Stratix 10 SoC, which combines ARMv8 cores with a tightly coupled FPGA fabric.

Three independent communication infrastructures enable an unparalleled combination of flexibility and performance: The M2DC server provides dedicated networks for monitoring and control, for management and compute, as well as for integrated highly flexible high-speed, low-latency communication between the microservers. Further increase in resource efficiency compared to other state-of-the-art solutions is targeted with the integration of system efficiency enhancements into the server architecture. At the hardware level, the SEEs are mainly realized using accelerators based on FPGAs, GPUs, or MPSoCs. At the software level, the accelerators will provide a wide variety of mechanisms for global system efficiency enhancements ranging from application-independent system-level functions via enhancements that support a complete class of applications to dedicated accelerators for a specific target application.

The M2DC middleware provides a data center capable abstraction of the underlying heterogeneity of the server based on OpenStack Ironic, providing bare metal (micro)server software deployment and lifecycle management. OpenStack Ironic will be extended toward handling the dynamic and heterogeneous nature of the microservers and hardware accelerators. Maintainability of the platform is ensured by a hardware/software infrastructure that is based on established standards, and advanced power and thermal management techniques, utilizing the rich sensorization of the server platform.

Since the developed microservers are based on established computer on module standards from the embedded domain, the realized platforms can be efficiently used not only for hyperscale data centers but also for edge computing in IoT and industrial IoT applications. Deploying the same hardware platforms at the edge and in the data center is seen as a promising feature to increase IoT programmer productivity and to decrease overall TCO.

Acknowledgements This work was supported in part by the European Union's Horizon 2020 research and innovation program, under grant no. 688201, Modular Microserver DataCentre (M2DC).

References

1. Agosta G, Barenghi A, Federico AD, Pelosi G (2015) Opencl performance portability for general-purpose computation on graphics processor units: an exploration on cryptographic primitives. Concur Comput Prac Exp 27(14):3633–3660. https://doi.org/10.1002/cpe.3358
2. Agosta G, Barenghi A, Pelosi G (2012) Exploiting bit-level parallelism in GPGPUs: a case study on KeeLoq exhaustive search attacks. In: Mühl G, Richling J, Herkersdorf A (eds) ARCS 2012 workshops, 28 Feb–2 Mar 2012. München, Germany, LNI, GI, vol 200, pp 385–396
3. Agosta G, Barenghi A, Santis FD, Pelosi G (2010) Record setting software implementation of DES using CUDA. In: Latifi S (ed) Seventh international conference on information technology: new generations, ITNG 2010. Las Vegas, Nevada, USA, 12–14 Apr 2010, pp 748–755. IEEE Computer Society. https://doi.org/10.1109/ITNG.2010.43
4. Amunts K, Lindner A, Zilles K (2014) The human brain project: neuroscience perspectives and German contributions. e-Neuroforum 5(2):43–50
5. Augonnet C, Thibault S, Namyst R, Wacrenier PA (2011) StarPU: A Unified Platform for Task Scheduling on Heterogeneous Multicore Architectures. Concurr. Comput. : Pract. Exper. 23(2):187–198. https://doi.org/10.1002/cpe.1631
6. vor dem Berge M, Christmann W, Volk E, Wesner S, Oleksiak A, Piontek T, Costa GD, Pierson JM (2012) CoolEmAll—models and tools for optimization of data center energy-efficiency. In: Sustainable internet and ICT for sustainability (SustainIT), 2012, pp 1–5
7. Ciznicki M, Kopta P, Kulczewski M, Kurowski K, Gepner P (2013) Elliptic solver performance evaluation on modern hardware architectures. In: Parallel processing and applied mathematics, pp 155–165. Springer
8. Ciznicki M, Kulczewski M, Kopta P, Kurowski K (2016) Scaling the GCR solver using a high-level stencil framework on multi-and many-core architectures. In: Parallel processing and applied mathematics, pp 594–606. Springer
9. Durand Y, Carpenter PM, Adami S, Bilas A, Dutoit D, Farcy A, Gaydadjiev G, Goodacre J, Katevenis M, Marazakis M, Matus E, Mavroidis I, Thomson J (2014) EUROSERVER: energy efficient node for european micro-servers. In: 2014 17th Euromicro conference on digital system design (DSD), pp 206–213. https://doi.org/10.1109/DSD.2014.15
10. Eicker N (2015) Taming heterogeneity by segregation—the DEEP and DEEP-ER take on heterogeneous cluster architectures. http://hdl.handle.net/2128/9379
11. Griessl R, Peykanu M, Hagemeyer J, Porrmann M, Krupop S, vor dem Berge M, Kiesel T, Christmann W (2014) A scalable server architecture for next-generation heterogeneous compute clusters. In: Proceedings of the 2014 12th IEEE international conference on embedded and ubiquitous computing, EUC '14, pp 146–153. IEEE Computer Society, Washington, DC, USA. https://doi.org/10.1109/EUC.2014.29
12. Lachmair J, Merényi E, Porrmann M, Rückert U (2013) A reconfigurable neuroprocessor for self-organizing feature maps. Neurocomputing 112(SI). https://doi.org/10.1016/j.neucom.2012.11.045
13. Lhuillier Y, Philippe JM, Guerre A, Kierzynka M, Oleksiak A (2014) Parallel architecture benchmarking: from embedded computing to HPC, a FIPS project perspective. In: Proceedings of the 2014 12th IEEE international conference on embedded and ubiquitous computing, EUC '14, pp 154–161. IEEE Computer Society, Washington, DC, USA. https://doi.org/10.1109/EUC.2014.30
14. Oleksiak A, Kierzynka M, Agosta G, Brandolese C, Fornaciari W, Pelosi G et al (2016) Data centres for IoT applications: the M2DC approach (Invited Paper). In: IEEE international

conference on embedded computer systems: architectures, modeling, and simulation (IC-SAMOS 2016). IEEE
15. OpenStack Technical Committee: OpenStack Ironic—Baremetal Provisioning. https://wiki.openstack.org/wiki/Ironic. Accessed 1 June 2017
16. PICMG: PICMG COM.0 R2.1—Com express module base specification. http://www.picmg.org. Accessed 1 June 2017
17. Rajovic N, Carpenter PM, Gelado I, Puzovic N, Ramirez A, Valero M (2013) Supercomputing with commodity CPUs: are mobile SoCs ready for HPC? In: Proceedings of the international conference on high performance computing, networking, storage and analysis, SC '13, pp 40:1–40:12. ACM, New York, NY, USA. https://doi.org/10.1145/2503210.2503281
18. Rojek KA, Ciznicki M, Rosa B, Kopta P, Kulczewski M, Kurowski K, Piotrowski ZP, Szustak L, Wojcik DK, Wyrzykowski R (2015) Adaptation of fluid model eulag to graphics processing unit architecture. Concurrency and Computation: Practice and Experience 27(4):937–957
19. Toradex: Apalis Computer Module—Module Specification. http://developer.toradex.com/hardware-resources/arm-family/apalis-module-architecture. Accessed 1 June 2017

Chapter 7
Towards an Energy-Aware Framework for Application Development and Execution in Heterogeneous Parallel Architectures

Karim Djemame, Richard Kavanagh, Vasilios Kelefouras, Adrià Aguilà, Jorge Ejarque, Rosa M. Badia, David García Pérez, Clara Pezuela, Jean-Christophe Deprez, Lotfi Guedria, Renaud De Landtsheer and Yiannis Georgiou

7.1 Introduction

The emergence of new applications (as well business models) in the Internet of Things (IoT), Cyber-Physical Systems (CPS), embedded systems, cloud and edge computing domains are transforming the way we live and work [1].

As the range of these applications continues to grow, there is an urgent need to design more flexible software abstractions and improved system architectures to

K. Djemame · R. Kavanagh · V. Kelefouras
School of Computing, University of Leeds, Leeds, UK
e-mail: K.Djemame@leeds.ac.uk

R. Kavanagh
e-mail: R.E.Kavanagh@leeds.ac.uk

V. Kelefouras
e-mail: V.Kelefouras@leeds.ac.uk

A. Aguilà · J. Ejarque · R. M. Badia
Barcelona Supercomputing Center (BSC),
Barcelona, Spain
e-mail: adria.aguila@bsc.es

J. Ejarque
e-mail: jorge.ejarque@bsc.es

R. M. Badia
e-mail: rosa.m.badia@bsc.es

D. García Pérez · C. Pezuela (✉)
Atos Research & Innovation,
Atos Spain SA, Madrid, Spain
e-mail: clara.pezuela@atos.net

D. García Pérez
e-mail: david.garciaperez@atos.net

C. Kachris et al. (eds.), *Hardware Accelerators in Data Centers*,
https://doi.org/10.1007/978-3-319-92792-3_7

fully exploit the benefits of the heterogeneous architectures on which they operate, e.g. CPU, GPU, heterogeneous CPU+GPU chips, FPGA and heterogeneous multi-processor clusters all of which with various memory hierarchies, sizes and access performance properties. In addition to showcasing such achievement, part of the process requires opening up the technologies to an even broader base of users, making it possible for less specialised programming environments to use effectively hiding complexity through novel programming models. Therefore, software plays an important role in this context.

On the other hand, computer systems have faced significant power consumption challenges over the past 20 years. These challenges have shifted from the devices and circuits level, to their current position as first-order constraints for system architects and software developers. A common theme is the need for low-power computing systems that are fully interconnected, self-aware, context-aware and self-optimising within application boundaries [2]. Thus, power saving, performance and fast computational speed are key requirements in the development of applications such as IoT and related computing solutions.

The project Transparent heterogeneous hardware Architecture deployment for eNergy Gain in Operation (TANGO) aims to simplify the way developers approach the development of next-generation applications based on heterogeneous hardware architectures, configurations and software systems including heterogeneous clusters, chips and programmable logic devices. The chapter will therefore present: (1) the incorporation of a novel approach that combines energy awareness related to heterogeneous parallel architectures with the principles of requirements engineering and design modelling for self-adaptive software-intensive systems. This way, the energy efficiency of both heterogeneous infrastructures and software is considered in the application development and operation lifecycle, and (2) an energy efficiency-aware system architecture, its components, and their roles to support key requirements in the environment where it runs such as performance, time-criticality, dependability, data movement, security and cost-effectiveness.

The remainder of the chapter is structured as follows: Sect. 7.2 describes the proposed architecture to support energy awareness. Sections 7.3–7.5 discuss key architectural components and their role to enact optimal, in terms of requirements

J.-C. Deprez · L. Guedria · R. De Landtsheer
CETIC, Charleroi, Belgium
e-mail: jean-christophe.deprez@cetic.be

L. Guedria
e-mail: lotfi.guedria@cetic.be

R. De Landtsheer
e-mail: renaud.delandtsheer@cetic.be

Y. Georgiou
Bull Atos Technologies, Les Clayes-sous-Bois, France
e-mail: yiannis.georgiou@atos.net

and Key Performance Indicators (KPIs), application construction, deployment and operation, respectively. Section 7.6 presents related work. In conclusion, Sect. 7.7 provides a summary of the research and plans for future work.

7.2 System Architecture

The high-level architecture is introduced on a per component basis, as shown in Fig. 7.1. Its aim is to control and abstract underlying heterogeneous hardware architectures, configurations and software systems including heterogeneous clusters, chips and programmable logic devices while providing tools to optimise various dimensions of software design and operations (energy efficiency, performance, data movement and location, cost, time-criticality, security and dependability on target architectures).

Next, the architecture is discussed in the context of the application life cycle: construction, deployment and operation. It is separated into remote processing capabilities in the upper layers, which in turn is separated into distinct blocks that support the standard application deployment model (construct, deploy, run, monitor and adapt) and local processing capabilities in the lowest layer. This illustrates support for secure embedded management of IoT devices and associated I/O.

The first block, *Integrated Development Environment (IDE)*, is a collection of components to facilitate the modelling, design and construction of applications. The components aid in evaluating power consumption of an application during its construction. A number of plugins are provided for a front-end IDE as a means for developers to interact with components within this layer. Lastly, this layer enables architecture agnostic deployment of the constructed application, while also maintaining low-power consumption awareness. The components in this block are as follows: (1) *Requirements and Design Tooling*: aims at guiding the development and configuration of applications to determine what can be targeted in terms of Quality of Service (QoS), Quality of Protection (QoP), cost of operation and power consumption behaviour when exploiting the potential of the underlying heterogeneous hardware devices; (2) *Programming model* (*PM*): supports developers when coding their applications. Although complex applications are often written in a sequential fashion without clearly identified APIs, the PM let programmers annotate their programs in such a way that the programming model runtime can then execute them in parallel on heterogeneous parallel architectures. At runtime, applications described for execution with the programming model runtime are aware of the power consumption of components implementation; and (3) *Code Profiler*: plays an essential role in the reduction of energy consumed by an application. This is achieved through the adaptation of the software development process and by providing software developers the ability to directly understand the energy footprint of the code they write. The proposed novelty of this component is in its generic code-based static analysis and energy profiling capabilities (Java, C, C++, etc. available in the discipline of

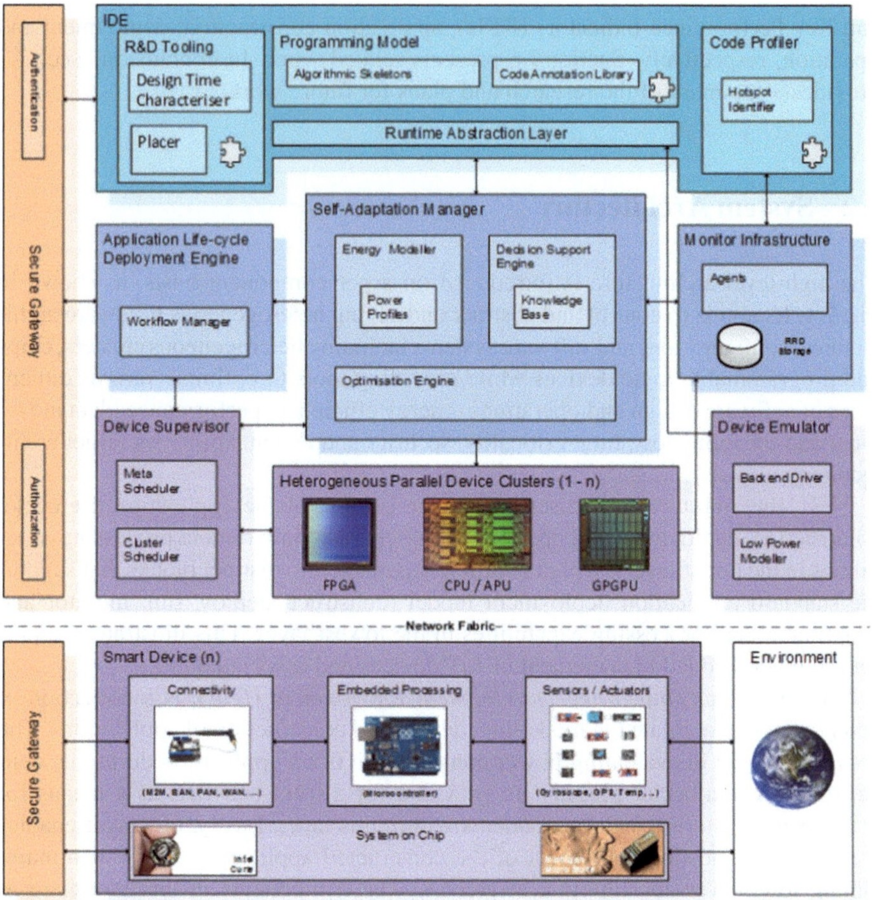

Fig. 7.1 Reference architecture

mobile computing) that enables the energy assessment of code out-of-band of an application's normal operation within a developer's IDE.

The second block consists of a set of components to handle the placement of an application considering energy models on target heterogeneous parallel architectures. It aggregates the tools that are able to assess and predict performance and energy consumption of an application. Application-level monitoring is also accommodated, in addition to support self-adaptation for the purpose of making decisions using application-level objectives given the current state of the application. The components in this block are as follows: (1) *Application Life cycle Deployment Engine*: this component manages the life cycle of an application deployed by the IDE. Once a deployment request is received, this component must choose the infrastructure that is most suitable according to various criteria, e.g. energy constraints/goals that indicate the minimum energy efficiency that is required/desired for the deployment and operation of an application; (2) *Monitor Infrastructure*: this component is able to monitor the heterogeneous parallel devices (CPU, memory, network, etc.) that are being consumed by a given application by providing historical statistics for device metrics. The monitoring of an application must be performed in terms of power/energy consumed (e.g. Watts that an application requires during a given period of its execution) and performance (e.g. CPU that an application is consuming during a given period of its execution); (3) *Self-adaptation Manager*: This component provides key functionality to manage the entire adaptation strategy applied to applications and Heterogeneous Parallel Devices (HPDs). This entails the dynamic optimisation of energy efficiency, time-criticality, data movement and cost-effectiveness through continuous feedback to other components within the architecture and a set of architecture specific actuators that enable environmental change. Examples of such actuators could be redeployment to another HPD, restructuring a workflow task graph or dynamic recompilation. Furthermore, the component provides functionality to guide the deployment of an application to a specific HPD through predictive energy modelling capabilities and policies, defined within a decision support engine, which specify cost constraints via Business-Level Objectives (BLOs).

The last block above the network fabric line addresses the heterogeneous parallel devices and their management. The application admission, allocation and management of HPDs are performed through the orchestration of a number of components. Power consumption is monitored, estimated and optimised using translated application-level metrics. These metrics are gathered via a monitoring infrastructure and a number of software probes. At runtime HPDs will be continually monitored to give continuous feedback to the self-adaptation manager. This will ensure the architecture adapts to changes in the current environment and in the demand for energy. Optimizations take into account several approaches, e.g. redeployment to another HPD, dynamic power management policies considering heterogeneous execution platforms and application energy models. The components in this block are as follows: (1) *Device Supervisor*: provides scheduling capabilities across devices during application deployment and operation. This covers the scheduling of workloads of both clusters (Macrolevel, including distributed network and data management) and HPDs (Microlevel, including memory hierarchy management). The component essentially realises abstract workload graphs, provided to it by the Application Life cycle Deployment Engine component, by mapping tasks to appropriate HPDs; (2) *Device Emulator (DE)*: is responsible for delivering the initial mapping of the application tasks onto the nodes/cores (at compile time), i.e. which application task should run on each node/core. The mapping procedure is static and thus it does not take into account any runtime constraints or runtime task mapping decisions. The TANGO user can choose between (a) a good solution in low time and (b) a (near)-optimum solution in a reasonable amount of time (depending on the application complexity and on the number of the available nodes/cores). Emulation of the application tasks on the HPDs is necessary in order to compute the corresponding performance and energy consumption values. The novelty of the DE component is that it reduces the number of different emulations required by order(s) of magnitude and therefore the time needed to map the tasks on the HPDs.

Furthermore, a secure gateway supports pervasive authentication and authorization, which at the core of the proposed architecture enables both mobility and dynamic security. This protects components and thus applications from unauthorised access, which in turn improves the dependability of the architecture as a whole.

7.3 Application Development

7.3.1 Design-Time Tooling

Designing and developing software with an efficient execution on distributed environment with fairly standard homogeneous processing devices is already a difficult exercise. This complexity explodes when targeting a heterogeneous environment composed not only of distributed multicore CPU nodes but also including accelerators with many-core CPUs, GPUs and FPGAs. In the current era of heterogeneous hardware, software development teams thus face the daunting task of designing software capable to exploit underlying heterogeneous hardware devices available to the most of their capability with the goal to achieve optimal runtime and energy performance.

The algorithmic decomposition chosen to solve the problem at hand and the selected granularity of computing task determine software execution efficiency on a given underlying hardware hence affect time and energy performance. For instance, many algorithms exist for matrix operations, data sorting or finding a shortest path in a graph. Developers already take into account data properties such as matrix sizes or degree of graph connectedness to select an algorithm with optimal time and energy performance. Nowadays, they must also consider capabilities offered by hardware in terms of parallel processing and data throughput. Such hardware capabilities influence design decision on algorithmic decomposition and task granularity choices to achieve efficient performance. For instance, time and energy performance associated with matrix multiplication on GPU or FPGA is directly influenced by matrix data sizes as well as the level of parallelism possible on a each different kind of processing nodes as well as their clock speed, their memory capacity, their data transfer latencies, internally within the chip and externally through their I/O interfaces. In other words, the most appropriate algorithmic decomposition and task granularity are jointly influenced by data properties as well as the capabilities of the underlying heterogeneous hardware available.

In addition to designing software for today's operational conditions, developers must strike the right balance between achieving an optimal performance now and keeping a design implementation flexible and evolvable for tomorrow's new hardware. The most efficient algorithmic decomposition and task granularity for today's heterogeneous hardware and dataset properties might evolve. In the worst case, evolution in hardware or data properties impacts software design and architecture forcing developers to adapt drastically the application code, that is, another algorithm must be implemented in order to better exploit the new hardware or the new kind of data. In less radical situations, a given overall software architecture and algorithms can remain unaltered. Only the task granularity must be adapted to process larger quantity of data at once for instance. New technologies and programming models such as OpenCL or OmpSs/COmpS [3, 4] can facilitate accommodating task granularity changes without much effort, hence keeping software implementation fairly evolvable. However, it still remains the job of developers to identify the appropriate task granularity for achieving improved time and energy performance and to provide this granularity information to the underlying technology or programming modelling tools. One of the goals of TANGO project is to provide design-time tooling to help developers to make insightful design decisions to implement their software so as to exploit the underlying hardware irrespective of the programming technologies and programming models chosen.

The initial approach to guide design decision, proposed in the first year of TANGO, relies on the rapid prototyping of the various simple software building blocks needed in a given application. The first step for developers consists of developing a set of simple prototypes for selected building blocks, for instance, for the different algorithms needed to solve multi-physics problems or to perform efficient image processing. Each prototype implements a particular algorithmic decomposition and task granularities for one of the identified simple software building blocks. For instance, a C or CUDA implementation of matrix multiplication will, respectively, target CPUs or NVIDIA GPUs nodes.

Developers can usually find alternative implementations of simple software building blocks that targets processors with fixed instruction set such as multicore, many-core and GPU. These implementations rely on programming technologies such as MPI, OpenMP, CUDA or OpenCL. On the other hand, the use of FPGA and other reconfigurable hardware has so far remained more complex and only used by much fewer experts. To address this issue, the TANGO development-time tooling proposes a tool, named Poroto, to ease porting segments of standard higher level code to FPGA. While OpenCL has recently proposed synthetisation for FPGA as part of its compilation toolchain, in many cases, developers only have implementation of simple building blocks in C code (or other programming languages). In such cases, an initial prototype implementation may be easier with Poroto than having to re-write the current C code in OpenCL. By annotating portions of C code with Poroto pragmas enables the generation of associated FPGA kernels and their interfacing to the code running on the CPU of the host machine through a PCI bus. The main processing program remains in C and is augmented with the necessary code, encapsulated in a C wrapper file that handles data transfer control to and from the offloaded FPGA computations. The C portion to be offloaded on FPGA is actually transformed into an equivalent VHDL program leveraging open-source C to VHDL compilers such as ROCCC, PandA or other HDL code generation tools. Subsequently, the VHDL can be passed to the lower level synthesis toolchains from the particular FPGA vendor like Xilinx or Intel/Altera to generate bitstream for a specific FPGA target. Concerning data transfer to/from the FPGA, Poroto currently relies on a proprietary technology.

However, an ongoing TANGO effort consists of replacing this proprietary technology with RIFFA (an open-source framework) to achieve similar data transfer operations.

Once the various prototypes of the different simple software building blocks have been implemented, compiled and deployed on the different targeted heterogeneous hardwares, it becomes possible to obtain benchmarks with different representative datasets on each of the prototype variants. The benchmarking exercise is not restricted to FPGA implementation, the initial code of simple software block can be executed on multicore and many-core CPUs and if code also exists for GPU, it can also be included in the benchmarking exercise.

After time and energy benchmarks for the different prototype implementations of the various simple blocks have been collected from the execution on the different heterogeneous hardwares targeted, developers must then identify an optimal way to place a combination of prototype implementations on the various hardware devices available in order to implement their complete solution. This optimisation problem between time and energy is not simple to solve in particular when considering different prototype implementations of several simple blocks competing for various heterogeneous hardware resources, and thus it becomes very useful to automate this optimisation exercise.

In TANGO, the development-time tooling relies on an open-source optimisation engine originated from operational research named OscaR to search optimal ways to map the implementation of different software blocks on the different heterogeneous hardware nodes. Specifically, the Placer finds optimal mappings of software component onto heterogeneous hardware, selects appropriate implementations of these tasks and performs software tasks scheduling for optimising energy performance while meeting specified timing constraints. Placer is implemented on top of the constraint programming engine of OscaR.

From an initial performance application design, it is then possible to further optimise application code by migrating from Poroto annotations to the COMPSs and OmpSs programming model in order to achieve concurrent execution of an algorithm on different heterogeneous processing nodes. This programming model is presented in the next subsection.

7.3.2 Programming Model

To manage the implementation of parallel applications for heterogeneous distributed computing environments, the TANGO programming model proposes the combination of two StarSs programming models and runtimes developed at Barcelona Supercomputing Center (BSC). StarSs is a family of task-based programming models where developers define some parts of the application as tasks indicating the direction of the data required by those tasks. Based on these annotations, the programming model runtime analyses data dependencies between the defined tasks, detecting the inherent parallelism and scheduling the tasks on the available computing resources, managing the required data transfers and performing the task execution. The StarSs family is currently composed of two frameworks: COMP superscalar (COMPSs) [3], which provides the programming model and runtime implementation for distributed platforms such as clusters, grids and clouds, and Omp Superscalar (OmpSs) [4], which provides the programming model and runtime implementation for shared memory environments such as multicore architectures and accelerators (such as GPUs and FPGAs).

In the case of TANGO, we propose to combine these programming models in a hierarchical way, where an application is mainly implemented by a workflow of coarse-grain tasks annotated with the COMPSs programming models. Each of these coarse-grain tasks can be implemented as a workflow of fine-grain tasks developed with OmpSs. At runtime, coarse-grain tasks will be managed by COMPSs runtime optimising the execution in a platform level by distributing tasks in the different compute nodes according to the task requirements and the cluster heterogeneity. On the other hand, fine-grain tasks will be managed by OmpSs which will optimise the execution of tasks in a node level by scheduling them in the different devices available on the assigned node.

This combination presents different advantages with respect to other approaches: First, it allows developers to implement parallel application in a distributed heterogeneous resources without changing the programming model paradigm. The programmer does not require programming model and APIs. It just requires to decide which parts are tasks, the direction of its data and its granularity. Second, developers do not have to deal with programming data movements like in MPI. The programming model will analyse data dependencies and keep track of the data locations during the execution. So, it will try to schedule tasks as close as data or transparently doing the required data transfer to exploit the maximum parallelism. Third, we have extended the versioning and constraints capabilities of these programming models. With these extensions, developers will be able to define different versions of tasks for different computing devices (CPU, GPUs and FPGA) or combinations of them. So, the same application will be able to adapt to the different capabilities of the heterogeneous platform without having to modify the application. During the execution, the programming model runtime will be in charge of optimising the execution of the available resources in a coordinated way. In platform scheduling, the runtime will schedule the task in the different compute node resources, deciding which task can run in parallel in each node and manage that the different tasks are not colliding in the use of resources by the affinity of task to devices. At the node level, the runtime is in charge of scheduling the fine-grain tasks in the resources assigned in the platform level scheduling.

7.3.2.1 Application Implementation Example

An example of how an application is implemented with TANGO programming model is shown next. This example implements a matrix multiplication by blocks in two levels. The first level splits the matrices into blocks and computes the matrix multiplication by block. Each block multiplication is defined as coarse-grain task. Each

```
int main(int argc, char **argv) {
  int N = atoi(argv[1]);
  int M = atoi(argv[2]);

  compss_on();

  cout << "Loading Matrices...\n";
  Matrix A = Matrix::init(N,Ml);
  Matrix B = Matrix::init(N,M);
  Matrix C = Matrix::init(N,M);

  cout << "Executing Multiplication...\n";
  for (int i=0; i<N; i++) {
    for (int j=0; j<N; j++) {
      for (int k=0; k<N; k++) {
        C.data[i][j]->multiplyBlocks(*A.data[i][k], *B.data[k][j]);
      }
    }
  }
  compss_off();
}
```

Fig. 7.2 Main workflow of the matrix multiplication

```
interface Matmul
{
  @Constraints(processors={@Processor(ProcessorType=CPU,
                                      ComputingUnits=4)});
  void Block::multiplyBlocks(in Block block1, in Block block2);

  @Constraints(processors={@Processor(ProcessorType=GPU,
                                      ComputingUnits=1)});
  @Implements(Block::multiplyBlocks);
  void Block::multiplyBlocks_GPU(in Block block1, in Block block2);
};
```

Fig. 7.3 Coarse-grain tasks definitions

matrix block can be decomposed in smaller blocks, and each block multiplication can be decomposed as a workflow of small block multiplications.

Figure 7.2 shows the main code of the benchmark application where a loop of the *multiply Blocks* coarse-grain tasks is implemented.

Figure 7.3 shows the interface file where the developer can define the methods which are defined as tasks. In this case, we have defined a task which has two implementations: one which runs in 4 CPU cores and another which runs in a GPU.

Finally, Fig. 7.4 depicts the implementation of the big block multiplication. In the first case, the fine-grain tasks are the computation of the different elements of the resultant matrix block. In the second case, the big matrix block is decomposed in smaller block in order to fit in the GPU device memory and fine-grain tasks are

```
void Block::multiplyBlocks(Block block1, Block block2) {
  for (int i=0; i<M; i++) {
    for (int j=0; j<M; j++) {
      #pragma omp task in(block1.data[i][0:M], \
                         block2.data[0:M][j]) out(data[i][j])
      for (int k=0; k<M; k++) {
         data[i][j] += block1.data[i][k]*block2.data[k][j];
      }
    }
  }
  #pragma omp taskwait
}

void Block::multiplyBlocks_GPU(Block block1, Block block2) {
  int NB = M/BSIZE;
  for (int i=0; i<NB; i++) {
    for (int j=0; j<NB; j++) {
      for (int k=0; k<NB; k++) {
        Muld(block1.data[i*NB+k], block2.data[k*NB+j],
             data[i*NB+j], NB);
      }
    }
  }
  #pragma omp taskwait
}

#pragma omp target device(cuda) ndrange(2, 64, 64, 32, 32)
#pragma omp task in(A[0:NB*NB], B[0:NB*NB]) inout(C[0:NB*NB])
__global__ void Muld(double* A, double* B, int wA, int wB,
                      double* C, int NB);
```

Fig. 7.4 Fine-grain tasks definitions

defined as the multiplication of these small blocks. The fine-grain task in this case is the CUDA kernel defined by the Muld function.

7.4 Application Deployment

The application deployment is taking care by the Application Lifecycle Deployment Engine (ALDE) that takes cares of the following tasks: provide the application development tools information about the possible targeted architectures; build the application for different configurations of heterogeneous hardware architectures and libraries; prepare the application packets for deployment also, if possible, deploy the application to the targeted testbed; and finally, if the connection with the device supervisor is possible, it will report and monitor the execution of the application to the user. These steps are explained in more detail next.

An installation of ALDE can register several testbeds that can have a TANGO device supervisor or not. If the testbed does not have a device supervisor, the user or administration needs to input the hardware heterogeneous characteristics of it: RAM, CPUs, GPUs, number of nodes, etc. If the testbed has a TANGO device supervisor, ALDE will automatically connect to the testbed and recollect the node hardware information. This information will be exposed to the application development tools so they can notify the application developer the testbed heterogeneous capabilities also, some of the development tools could use this information to determine which is the best testbed to run a given application.

The building process of the application will be done by ALDE compiling the application for different combinations of targeted heterogeneous architectures and libraries. The usage of tools is like EasyBuild [5] or Spack [6]. The different compilations could then be manually selected by the user of self-adaptation manager to deploy the optimal code for the given available resources by the device supervisor.

After the application is compiled, it needs to be packetized. The final packet format will depend on the targeted architecture. ALDE supports just submitting the application to the device supervisor by simple binaries (typical HPC scenario). It also supports the creation of containers based on Docker [7] or Singularity [8]; this is both targeted to HPC and embedded environments that allow containers as an application distribution system. Finally, it also supports the generation of ISO images to be installed into heterogeneous embedded devices.

If the targeted heterogeneous architecture has an online device supervisor, ALDE has the possibility to connect to it and monitor the execution of the application. During the third year of the project, in this case, it is also expected that ALDE would supervise the data transfer to the selected architecture for the execution of application.

7.5 Application Execution

7.5.1 Device Supervisor

The Device Supervisor (DS) is responsible for efficiently delivering the computing power of heterogeneous devices to the applications based on their needs. It provides the means to enable the execution of applications upon the platforms' resources. In particular, it offers a number of parameters that enable the fine specification and usage of different types of resources (CPUs, GPUs, Memory, etc.) and their constraints for the optimal execution of the applications. Furthermore, it enables task placement and isolation upon devices during application deployment and operation.

Besides the various features and parameters for single application execution, this component allows the usage of compute platform by multiple users where jobs may even compete for the same resources. Hence, its main intelligence relies on resource selection techniques to find the most adapted resources to schedule the users' jobs while keeping a high system utilisation and low fragmentation.

Within the TANGO framework, the DS can get inputs from the Application Lifecycle Deployment Engine to execute jobs under particular parameters and it can follow the execution of the application and return intermediate state or final results to the ALDE. Optimization criteria (such as power consumption) and environment state are provided as input by the self-adaptation manager and monitoring infrastructure components, respectively.

The device supervisor in TANGO is represented by Slurm [9] which is an open-source resource and job management system. Slurm performs workload management on five of the ten most powerful computers in the world of the Top 500 list[1], including the system ranked number two, Tianhe-2, which features 3,120,000 computing cores.

Slurm is specifically designed for the scalability requirements of state-of-the-art supercomputers. It is based upon a centralised server daemon, `slurmctld` also known as the controller, which communicates with client daemons `slurmd` running on each computing node. Users can request the controller for resources to execute interactive or batch applications, referred to as jobs. The controller dispatches the jobs on the available resources, whether full nodes or partial nodes, according to a configurable set of rules. The Slurm controller also features a modular architecture composed of plugins responsible for different actions and tasks, such as job prioritisation, resources selection, task placement or accounting.

Most resource and job management systems today do not handle heterogeneous resources efficiently. They provide a complete SPMD (Single Program Multiple Data) support but limited MPMD (Multiple Program Multiple Data) support. Limited MPMD support means that even if users can specify different binaries to be used within a parallel job, all the tasks are currently associated with the same resources requirements. To be able to leverage all the benefits of platforms with heterogeneous resources, we need to be able to specify different heterogeneous resources within the same job and be able to support the MPMD model. This support will enable users willing to harness different types of hardware resources inside the same MPI application, having part of their code run on GPUs, while other parts are executed on standard CPUs with specific low amount of memory and the last part on CPUs with large amount of memory. Currently, we are obliged to request the most complete set of resources for each task wasting some of the hardware with tasks that will not need all of them. In some cases, the total configuration required to run such a job does not even exist as all the nodes of the cluster may not provide all the hardware features.

Hence, the device supervisor component of Tango will be represented by an enhanced version of Slurm resource and job management system specifically designed to support heterogeneous resources.

[1] https://www.top500.org/list/2017/11/.

7.5.2 Energy Modeller

Energy modelling can be used at multiple phases of an application life cycle. At deployment time, it helps with the assignment of resources to an application and at runtime it aids a continuing energy mitigation strategy.

The Energy Modeller (EM) provides power and energy consumption information for compute devices in the current, future and historical contexts, thus providing key information that guides the selection of the most appropriate configuration of an application within a heterogeneous environment, with the aim of minimising energy consumption, acting as a key advisory component in the energy reduction process. It provides the mathematical models that estimate the power consumption and energy usage of a given deployment decision. Thus, it is able to advise and drive the selection of hardware for service deployment and advise the process of self-adaptation.

The energy modeller's facility to assess historic energy consumption forms the heart of any advisory service for end users who wish to understand the energy consumption of their application. The advice to end users goes further by informing them of the current power consumption of their software and hardware setup, and thus they can gauge the current impact of running their applications.

The energy modeller requires the use of models to determine from a host's resources usage the likely future energy consumption, as well as providing a means of attributing power consumption to a particular application.

An estimation of the power consumption of an application or physical resource derives from two aspects. The first is the correct profiling of the resources characteristics, encompassing aspects such as its idle energy consumption and energy consumption under various load conditions. The second aspect is the profiling of the workload to be performed. This workload derives from the application that is to be characterised based upon the hardware it runs upon. These two profiles combined therefore advance the understanding of how much energy a application is expected to consume in the future.

The aspect of correctly charactering resource takes care during the calibration process. The calibration process must provide repeatable conditions that generate a sequence of precise loads on the physical host undergoing measurement. The aim is to tightly control the environment while running an experiment to gain an accurate mapping between the resource utilisation and power consumption. This data can then be used as the basis of predicting future power consumption/energy usage, attributing power to a given workload as well as providing faster and more responsive measures of current power consumption especially for short runs of an application. The process of applying fixed calibration loads is illustrated in Fig. 7.5.

A sequence of runs (marked as (a)) is shown with increasing utilisation, with small gaps between each run. The duration (a) can be chosen based upon any averaging window of the reported sensor data. A longer time period (a) gives a greater chance of the reported utilisation and power level stabilising. Issues such as averaging, unsynchronised metrics, network delay or caching mechanisms can all have their effects on calibration accuracy. A key solution to this is to discard values at the start

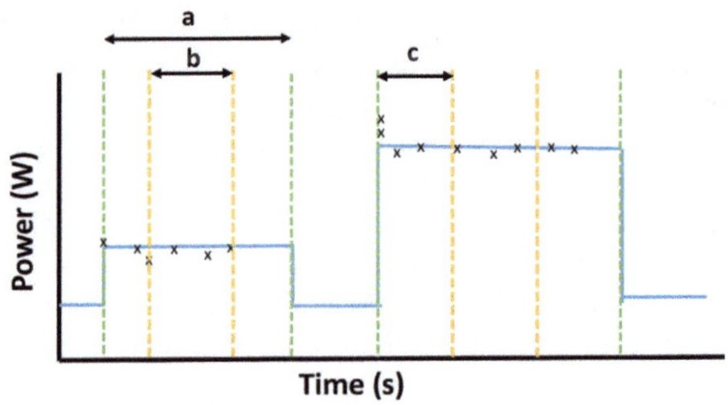

Fig. 7.5 The construction of artificial traces for calibration

and end of an experimental run (indicated by (c) in Fig. 7.5). In addition to this, it eliminates experimental error such as load spike above the intended target load when each run starts. The final set of datapoints in the area indicated by (b) represents the best calibration data. One advantage of a model is that once it is calibrated, even if the power measurement sensor reports an average value, an instantaneous estimated power consumption value can be obtained without averaging and at a higher temporal granularity through the model.

The second aspect of attributing power to a given application needs reasonable way to allocate power consumption. One such way is to consider the system's idle energy usage as well as any active power consumption, given a specific application's load. The idle energy/power consumption should be evenly distributed among the applications that are running upon the host machine. The remaining energy is then allocated based upon the induced load. This is described in Eq. (7.1) where EU_P_x is the application's power consumption and $Host\_P$ is the measured host power consumption. EU_Util_x is the application's CPU utilisation and EU_Count is the count of applications on the host machine. EU_Util_y is the CPU utilisation of a member of the set of applications on the named host. $Host\_Idle$ is the host's measured idle power consumption.

$$EU\_P_x = Host\_Idle + (Host_P - Host\_Idle) \times \frac{EU\_Util_x}{\sum_{y=1}^{EU\_Count} EU\_Util_y} \tag{7.1}$$

7.5.3 *Self-adaptation Manager*

The Self-Adaptation Manager (SAM) is the principle component in the middleware for coordinating self-adaptation. It plays an essential role in maintaining power, energy and performance and goals of an application at runtime. Its primary focus is upon providing the infrastructure runtime self-adaptation capabilities with a particular focus on trade-off management for the applications. This is achieved through the careful consideration of violations in service quality and the actuators that can be utilised to perform self-adaptation. This adaptation covers both macro- and micro-aspects of an application's deployment upon heterogeneous parallel architectures. The levels differ in that deployed applications may be submitted to a given node, but the node also has heterogeneity in that various accelerators might be utilised and configured for usage by the varying applications that are running.

The self-adaptation manager follows a MAPE [10] control loop pattern of monitor analyse, plan and execute. Adaptation in this cycle considers these main aspects:

- the varying level of QoS required, mainly either real-time high quality of service or best effort services;
- the various implementations of an application, which will have means to use various accelerators;
- the performance of each implementation on the accelerators (affinity towards an adaptor);
- the availability/demand for accelerators and resources;
- the malleability of an application;
- the required pace of response (how quickly change occurs and real-time requirements); and
- the acceptable frequency of adaptation (avoiding over adaptation).

These collectively will give an application an affinity towards various accelerators and application configurations. QoS in TANGO is formed of two distinct categories of application. The first is real-time applications that require a high level of quality of service, i.e. that they have priority to resources while best effort services are expected to comprise the rest of the tasks.

Applications are expected to have various different implementations, each of which will be able to be executed on only some of the accelerators. The availability of these variations will be important as it offers the possibility for the SAM to select and switch between actuators dependent upon their availability. The quality of each implementation varies and will depend upon if the implementation can be structured in a way that takes advantage of the accelerator. This will give varying degrees of speed up, which will give a notion of affinity of an application to a given adaptor. The adaptors are a limited valuable resource that is not in all cases shareable. Fine-grain pre-emption in NVIDIA GPUs has only become available in the Pascal architecture, which was released in mid-2016 [11]. Given the limited access to resources some applications may throughout their lifecycle be able to scale or shrink their resource usage.

A portion of jobs will be rigid and unable to change their resource requirements, while some others will be mouldable to the resources that are available at deployment. A further set will be malleable and will be able to dynamically change their resource requirements. This is particularly useful in regards to the availability of accelerators. Added to the limits of access to accelerators, some adaptations will be required to be completed with very limited delay, such as video processing. This places limits on the types of actuation that is permitted. This gives rise to a bias towards smaller control loops that handle specific QoS requirements. In cases of adaptation, there is an acceptable frequency by which actuation is allowed to occur, such as once per minute. This is an important factor as it ensures applications are not interfered with and are allowed to perform useful work.

7.6 Related Work

Computing nodes are incorporating different types of devices in order to be more efficient when computing different types of applications by accelerating the computation at lower power. However, this heterogeneity brings more complexity in the application development, each of these devices has their own programming language or API to spawn the computation in the different devices. For instance, FPGA are traditionally programmed with the VHDL language; and for deploying and running the computation, developers have to use the toolchain provided by the FPGA vendor. A similar problem happens with the general-purpose GPUs. NVIDIA offers the CUDA framework [12] for programming and running applications in its devices and other vendors offer similar frameworks to do the same.

Current research is focusing their efforts on reducing the complexity of programming these heterogeneous nodes, as well as providing portability between architectures allowing the reuse of the code for similar devices. One of these examples is OpenCL [13]. It was born with the ambition of providing a common programming interface for heterogeneous devices (including not only GPUs but also DSPs and FPGAs). With a syntax based on C, it has had a significant impact because the same code could be used in several accelerators. However, similar to CUDA, it requires the programmer to write specific code for the device handling, which reduces programmability. OpenACC [14] is another example of programming standard for parallel computing designed to simplify parallel programming of heterogeneous CPU/GPU systems. Based on directives, the programmer can annotate the code to indicate those parts that should be run in the heterogeneous device. The OpenMP standard [15] tackles the programmability issues in a similar way as OpenACC with regard to the heterogeneous devices and also considers many other aspects of parallelism which makes it a stronger option.

However, these solutions are just managing the heterogeneity inside a node. If the application requires to run in several nodes (e.g. big amount of data or large parallelism), solutions mentioned before must be combined with other frameworks which manage the spawning of processes and data movements between the different computing nodes. Developers can attempt to do it by hand using the TCP/IP and threading libraries, which require a lot of programming effort and skills or use one of the parallel distributed computing frameworks. One of these frameworks is MPI [16], which provides an API for interchanging data messages between the different processes for SPMD applications. Other options are PGAS programming models such as UPC [17], which allow to create a global address space and use shared memory programs in different nodes. Both options are working quite well-running Single Process Multiple Data (SPMD) application in homogeneous clusters interconnected with a very fast network and SPMD application. However, in heterogeneous environments distributed across different locations, they are to reaching good performance. For these reasons, we have proposed to combine COMPSs and OmpSs which have good results in managing heterogeneity at platform and node level, and its programmability relies on the same task-based paradigm.

Prior to runtime considerations, developers must provide application code tailored for executing on heterogeneous hardware. Whether relying on MPI, OpenACC, OpenMP, OpenCL, CUDA, or plain old C in application code, developers must craft their implementation, i.e. decompose their algorithms and identify granularity of subtask of these algorithms in order to exploit the available heterogeneous hardware devices optimally. While full development toolchains exist in open source for fast prototyping on standard multi- and many-core CPU [18] and on GPU [19, 20], no integrated toolchain is available in open source for fast prototyping offloading on FPGA. Poroto provides such an integration over existing open-source frameworks such as ROCCC for the high-level synthesis from C to VHDL. Future plans include support of PandA framework and also better integration with RIFFA for generic and portable handling of data transfers between the main CPU and the FPGA board. Regarding the optimal mapping of software component and scheduling of tasks using this software component, PREESM [18] and Silexica [21] both have studied the problem. However, they are not publishing their ad hoc algorithms. *Placer* is implemented on the top of the operational research optimisation framework OscaR, whose foundations have been validated in several industry grade projects and products. Thus, Placer only needs to implement additional problem-specific code. This allows for high flexibility to support various requirements such as power cap and DVFS, among others as well as better readability for verifying the correctness of the optimisation search algorithm.

Resource management and job scheduling in traditional HPC systems are being performed by specialised software called RJMS. This software holds an important position in the HPC stack since it stands between the user workloads (jobs) and the hardware platform (resources). It is responsible for delivering computing power to applications efficiently. More than 2 decades of research and developments in the field has resulted in various open-source and proprietary versions of RJMS that exist today [22–26] offering basic and advanced functionalities to deal with HPC specialised platforms and workloads.

Since 2010, some newer generation schedulers such as Mesos [27] and Yarn [28] can execute both compute and data-intensive workloads based on new types of internal architectures trying to deal with scalability, efficiency and fault-tolerance issues. In this group, we can also add Flux [29] which is currently under active development and destined for extreme-scale HPC systems.

7.7 Conclusions

This chapter has highlighted the importance of providing novel methods and tools to support software developers aiming to optimise energy efficiency resulting from designing, developing, deploying and running software on HPAs while maintaining other quality aspects of software to adequate and agreed levels.

The specification of a proposed architecture has been presented, which includes the architectural roles and scope of the components. This architecture complies with standard HPAs and supports an IDE, an application deployment on HPA environments and heterogeneous parallel device environments. The design of the various architectural components was described, with emphasis on the requirements in order to support energy efficiency management, which is addressed during the complete life cycle of an application.

Future work includes the implementation of capabilities to perform continuous autonomic self-adaptation during runtime. This leverages fine-grained monitored metrics of heterogeneous parallel devices and application software to create an adaptation plan supporting the performance and cost goals of an application. It is achieved through advances in modelling and prototyping that enable power, cost and performance awareness during operation through emulation and simulation under various 'what-if' scenarios.

Acknowledgements This work has been supported by the European Commission through the Horizon 2020 Research and Innovation program under contract 687584 (TANGO project) by the Spanish Government under contract TIN2015-65316 and grant SEV-2015-0493 (Severo Ochoa Program) and by Generalitat de Catalunya under contracts 2014-SGR-1051 and 2014-SGR-1272.

References

1. Iot's challenges and opportunities (2017) A gartner trend insight report, Apr 2017
2. Djemame K, Armstrong D, Kavanagh RE, Deprez JC, Ferrer AL, Perez DG, Badia RM, Sirvent R, Ejarque J, Georgiou Y (2016) Tango: transparent heterogeneous hardware architecture deployment for energy gain in operation. In: Proceedings of the first workshop on program transformation for programmability in heterogeneous architectures, arXiv:1603.01407
3. Badia RM, Conejero J, Diaz C, Ejarque J, Lezzi D, Lordan F, Ramon-Cortes C, Sirvent R (2015) Comp superscalar, an interoperable programming framework. SoftwareX 3:32–36
4. Duran A, Ayguadé E, Badia RM, Labarta J, Martinell L, Martorell X, Planas J (2011) Ompss: a proposal for programming heterogeneous multi-core architectures. Parallel Process Lett 21(02):173–193
5. HPC UGent (2017) Easybuild: building software with ease. https://easybuilders.github.io/easybuild/
6. Lawrence Livermore National Laboratory (2017) Spack—package management tool
7. Docker Inc. (2017) Docker—a better way to build apps. https://www.docker.com/
8. Singularity (2017). https://singularity.lbl.gov/
9. Yoo AB, Jette MA, Grondona M (2003) Slurm: simple linux utility for resource management. In: Job scheduling strategies for parallel processing, pp 44–60
10. IBM (2005) An architectural blueprint for autonomic computing

11. Smith R (2016) Preemption improved: fine-grained preemption for time-critical tasks
12. NVIDIA Corp (2017) CUDA homepage. http://www.nvidia.es/object/cuda_home_new.htm. Accessed 3 May 2017
13. Stone JE, Gohara D, Shi G (2010) Opencl: a parallel programming standard for heterogeneous computing systems. Comput Sci Eng 12(3):66–73
14. OpenACC Application Programming Interface Specification (2017). http://www.openacc.org/specification. Accessed 3 May 2017
15. OpenMP Architecture Review Board (2017) OpenMP application programming interface specification. http://www.openmp.org/specifications/. Accessed 3 May 2017
16. MPI forum (2017) Message passing interface specification. http://mpi-forum.org/. Accessed 3 May 2017
17. Tarek A (2006) El-Ghazawi and Lauren Smith. Upc: unified parallel c. In: Proceedings of the 2006 ACM/IEEE conference on Supercomputing, pp 27. ACM
18. Pelcat M, Desnos K, Heulot J, Guy C, Nezan JE, Aridhi S (2014) Preesm: a dataflow-based rapid prototyping framework for simplifying multicore dsp programming. In: 2014 6th European embedded design in education and research conference (EDERC), Sept 2014, pp 36–40
19. GPU Open Consortium (2017) Code XL. http://gpuopen.com/compute-product/codexl/. Accessed 17 May 2017
20. NVIDIA (2017) NVIDIA CUDA toolkit. https://developer.nvidia.com/cuda-toolkit. Accessed 17 May 2017
21. Silexica GmbH (2017) Software design for multicore. https://silexica.com/. Accessed 17 May 2017
22. Capit N, Da Costa G, Georgiou Y, Huard G, Martin C, Mounié G, Neyron P, Richard O (2005) A batch scheduler with high level components. In: 5th international symposium on cluster computing and the grid (CCGrid 2005), Cardiff, UK, 9–12 May 2005, pp 776–783
23. Litzkow MJ, Livny M, Mutka MW (1988) Condor—a hunter of idle workstations. In: Proceedings of the 8th international conference on distributed computing systems, San Jose, California, USA, 13–17 June 1988, pp 104–111
24. Zhou S, Zheng X, Wang J, Delisle P (1993) Utopia: a load sharing facility for large, heterogeneous distributed computer systems. Softw Pract Exp 23(12):1305–1336
25. Adaptive Computing (2017) Moab HPC basic edition. http://www.adaptivecomputing.com/products/hpcproducts/moab-hpc-basic-edition/
26. Altair (2017) PBS professional open source project. http://www.pbspro.org/
27. Hindman B, Konwinski A, Zaharia M, Ghodsi A, Joseph AD, Katz RH, Shenker S, Stoica I (2011) Mesos: a platform for fine-grained resource sharing in the data center. In: Proceedings of the 8th USENIX symposium on networked systems design and implementation, NSDI 2011, Boston, MA, USA, 30 Mar–1 Apr 2011
28. Vavilapalli VK, Murthy AC, Douglas C, Agarwal S, Konar M, Evans R, Graves T, Lowe J, Shah H, Seth S, Saha B, Curino C, O'Malley O, Radia S, Reed B, Baldeschwieler E (2013) Apache hadoop YARN: yet another resource negotiator. In: ACM Symposium on Cloud Computing, SOCC '13, Santa Clara, CA, USA, 1–3 Oct 2013, pp 5:1–5:16
29. Ahn DH, Garlick J, Grondona M, Lipari D, Springmeyer B, Schulz M (2014) Flux: a next-generation resource management framework for large HPC centers. In: 43rd international conference on parallel processing workshops, ICPPW 2014, Minneapolis, MN, USA, 9–12 Sept 2014, pp 9–17

Chapter 8
Enabling Virtualized Programmable Logic Resources at the Edge and the Cloud

Kimon Karras, Orthodoxos Kipouridis, Nick Zotos, Evangelos Markakis and George Bogdos

8.1 Enhancing Compute Through Programmable Logic

Continuous, unabated performance increases are an innate part of the computer world from the invention of the first microprocessor until today. This trend appears to be, however, slowly but steadily coming to an end. Single thread performance gains have been slow to extract from each new technology node for more than 10 years. This was countered successfully by a massive push into parallel computing in the form of multi-core, multi-threaded processors which continues until the present, albeit while showing clear signs of running out of steam. The reason is that parallelism isn't a silver bullet that can be applied to every problem in infinite amount. Many common applications can't be parallelized further, and thus, any gains that can be extracted by increasing the number of cores and/or threads are limited to only small subset of all applications.

This has led academia and industry alike in search of a solution that will allow for the processing of the massive heaps of data that are being collected throughout the world by sensors, cameras, the Internet, etc. The most promising fruit of that research has by far been heterogeneous compute, a concept that encompasses anything from

K. Karras (✉) · O. Kipouridis · N. Zotos · G. Bogdos
Future Intelligence Ltd., London, UK
e-mail: kkarras@f-in.co.uk

O. Kipouridis
e-mail: akip@f-in.co.uk

N. Zotos
e-mail: nzotos@f-in.co.uk

G. Bogdos
e-mail: gbodgos@f-in.co.uk

E. Markakis
Technological Education Institute of Crete, Heraklion, Greece
e-mail: markakis@pasiphae.eu

C. Kachris et al. (eds.), *Hardware Accelerators in Data Centers*,
https://doi.org/10.1007/978-3-319-92792-3_8

CPUs with different core types on the same processor die to modern SoCs that include CPUs, GPUs and custom accelerators or custom processors like Google's Tensor Processing Unit [8] or Micron's Automata Processor [13].

These solutions are trying to address the processing of massive data, principally in the cloud but also increasingly at the edge, where a big part of that data is being generated. Both of these fields are ripe for exploration where a multitude of concepts are vying for domination. Regardless of the details of each architecture, the basic premise and ingredients are always the same, as are the issues plaguing the current cloud-centric paradigm which they're trying to solve, namely:

- Heaps of data being directed to the cloud for processing and/or storage, leading to congestion at the network leading to and from the data centers.
- Unacceptably high decision-making latency leading to delays in taking action based on collected data.

The former is a direct result of the centralized architecture, which treats edge nodes as simple data collection points and/or actuators and ships all the data to one or more massive data centers for analysis. As the number of edge devices explodes and the data they collect diversifies, the traffic being fed into data centers will be strained to the extreme, which means that either the network will become congested or unacceptably high investment into infrastructure will be required to avoid this. Congestion on the path to and from the data center will unavoidably lead to the second issue, latency. Decision-making latency means the time it takes to reach a decision regarding an action to be taken and implement that action wherever required, usually at the networks edge. Currently in order for this to happen, data needs to be sent up to the cloud, sifted through and the decision needs to propagated downward to the edge again over multiple, potentially unreliable hops. Thus, it is clear that there are tangible benefits to accelerating tasks into the cloud but also to shifting parts of the processing load toward the networks edge. This translates into increased processing requirements for small, low power mist and fog nodes, which are currently not up to the task and which can't simply use the powerful CPUs found in contemporary data centers due to power constraints. This makes the case for compute acceleration that delivers significant performance gains at low power consumption at the edge even more prescient than in the cloud.

One alternative that can offer impressive performance increase while maintaining very low power consumption, a feature critical in power-strapped edge nodes, are FPGAs. There's a long literature of examples where FPGAs provide impressive performance benefits in comparison to standard CPUs, which Sirowy and Fiorin sum up excellently [11]. At the same time, FPGAs can be reprogrammed quickly and infinitely so the tasks assigned to them can vary over time making them easily adaptable to changing demands and capable of receiving updated dynamically in the field. That being said, programmable logic has been marred by several caveats that hinder its adoption:

- While FPGAs are reprogrammable, the process does not allow for automated, remote deployment of tasks from a distance. Typically, programming has to be performed with a cable attached to the device.

- FPGA development requires very esoteric knowledge of the target device including low-level details like pin assignments and interface standards.

  This chapter proposed the Programmable Compute Platform (PCP) which addresses and alleviates both of these concerns by offering:
- A system consisting of both HW and SW that allows for the integration of an FPGA into the edge and/or cloud and the remote (re)deployment of tasks to it over a standard SW stack.
- A working hardware system with strictly defined, intuitive interfaces where tasks can work in a plug-and-play manner thus reducing the overhead to develop tasks for remote deployment.

In the current iteration of the platform, this is accomplished by utilizing a novel device called an FPGA SoC. We use this device to achieve an optimal division of labor between the A9 CPUs and the FPGA, with the CPUs executing the SW that enables the remote deployment of an HW task (similar to an HW virtual machine) on the programmable logic. This includes an agent that can reply to queries about the devices state and can receive a HW VM image and deploy it to the HW and/or create virtual network interfaces to integrate the device seamlessly into service function chains.

Our system is at the stage of a functioning proof of concept with all of the mandatory HW and SW components up and running on the Xilinx ZC706 development board. An h264 decoder is used as a test HW VM to highlight the benefits of using an FPGA as an accelerator and the performance and power advantages it brings.

The remainder of this chapter is organized as follows: Section 8.2 provides an overview of existing efforts in this area. Section 8.3 introduces the programmable cloud platform and its components, while Sect. 8.3.3 explains the changes made to the OpenStack controller to make it FPGA-aware. Section 8.4 provides preliminary results both for the platform and the HW VM being tested, and finally, Sect. 8.5 summarizes our findings and provides pointers for future work.

8.2 Prior Efforts in Virtualizing Programmable Logic

Research in integrating FPGAs in the existing cloud infrastructure is scarce though the topic has been gaining traction as of late. Edge acceleration has on the other hand been flying completely under the radar so far, even at a conceptual level. On the cloud front, the handful of research papers can be found on the subject is summarized here, with all of them converging on a single approach:

- Extending OpenStack Nova to allow for the deployment of FPGA-based systems.
- Provision of a static area in the FPGA to allow for the deployment of the NFs.
- A dynamically reconfigurable area in the FPGA in which the user will be able to deploy his NFs.

More specifically, Buma et al. [3] introduce a framework for deploying the so-called Virtualized FPGA Resources (VFRs) to an FPGA which focuses on the HW side of the problem. They take advantage of the OpenStack service called glance to deploy VM packages which contain FPGA bitstreams to the device. These are then picked up by an agent (running in SW on the compute node), which unpacks the VM and deploys the bitstream to the device. The advantage of this approach is that the FPGA is completely abstracted from OpenStack. Deployment of a VM from its perspective remains does not change in the slightest since the whole process is handled in the agent on the compute node. The agent manages the FPGA, which is divided into a static and a dynamic part. The static part contains all the modules necessary for the deployment of the VFRs and for the communication between the external world and the VFRs (including a network and a DRAM interface). It also performs the necessary handshaking between the VFR and the agent to start and stop it when deploying or retracting it. The dynamic area is where the VFRs are actually programmed. A NetFPGA-10G board is used for a simple prototype, which uses a load balancer application to demonstrate primarily the feasibility of the concept, but also the performance gains that can be reaped by migrating from an SW VM to an FPGA-based one.

Another similar if somewhat more comprehensive effort was published by Chen et al. [4]. This work includes many common elements with the system to be developed within this project. More specifically, it uses adapted OpenStack to deploy HW VM images to the FPGA in which it implements a layered architecture that enables the programmable logic in the FPGA to run these HW VMs in isolation. On the HW level, the programmable logic can be both space- and time-shared by introducing the notion of a segregated accelerator region in which the HW VMs are slotted in. Additionally, a static area is reserved on the programmable logic, which regulates access to DMA resources among the accelerators and manages jobs in them. On top of this HW infrastructure, there are three SW layers that facilitate the deployment and control of the HW VMs from host SW whether that is the cloud controller or SW running locally on the processor. The main limitation of this work is the limited connectivity offered to the accelerators (only a DMA connection to DRAM) and the fact that there is no provision for HW/SW VMs. Furthermore, the paper does not go into detail about how OpenStack had to be extended to support FPGA-based VMs.

A third approach for deploying VNFs to FPGA-based systems was published by Ge et al. [7]. This work shares a lot of commonalities with the [2] in that the programmable logic is divided into slots for accelerators to which tasks are dispatched. The accelerators are slotted in using partial reconfiguration which is handled by a VM running on standard x86 processor. Additionally, each accelerator is deployed over its own SW VM via OpenStack. This means that having an SW component is mandatory even if its just a shell for deploying an accelerator. In the system described it appears that both OpenStack and VMs are not tightly integrated into this system, meaning that the heterogeneity of the system is hidden from these components.

Another indispensable element of any virtualized system is the Hypervisor. Of particular interest for this work are potential Hypervisors that can be executed on the ARM A9 processors found on an FPGA SoC. One possible alternative was presented by Pham et al. [10]. The system introduces in this paper encompasses a wider system than just the Hypervisor. That system is based on the time-multiplexing of tasks on the programmable logic to accelerate SW executed on the Hypervisor. To accomplish this it creates an overlay onto the programmable fabric to allow for coarse-grained core design and deployment. The most interesting part of this work is the Hypervisor which runs on the ARM A9 cores and is based on the Codezero Hypervisor from Bell Labs. This could be reused in our effort, granted that the source code is available; however, modifications will most certainly have to be made to:

- Allow it to run on the ARM A9s found on the Zynq.
- Allow it to virtualize the FPGA resource.
- Communicate in a compatible manner with OpenStack Nova.

The effort required by each of these changes remains to be evaluated.

It's clear from the short overview provided in this section that while all existing solutions appear to agree on the basic features required for FPGA cloud integration, all of the platforms end up leaving something to be desired. PCP rectifies this by ticking all the boxes and offering a straightforward, intuitive FPGA-based device for the cloud end the edge.

8.3 A Virtualizable, Remotely Manageable FPGA SoC Platform

This section provides a detailed overview of the FPGA SoC-based programmable cloud platform. It consists of a hardware and a software component that together allows for the remote deployment of hardware tasks on the FPGA by using standard cloud management software, namely OpenStack. The hardware component is further elaborated upon in Sect. 8.3.1. The software running on the FPGA SoC's ARM CPUs allows the integration of the platform into OpenStack and is explained in Sect. 8.3.2. Finally, the changes that were performed in the OpenStack controller in order to allow it to utilize programmable logic-based devices are highlighted in Sect. 8.3.3.

PCP makes some concessions in order to simplify the design and strike an optimal balance between design complexity (and thus the overhead of managing the VMs' deployment) and the flexibility offered to the user. The first of the two decisions that were made were to only allow one HW VM to be deployed on the programmable fabric at any given time. This was done to avoid the complexities of parallel deployment of HW tasks on the FPGA. These have been discussed in detail in the past [1, 2, 9] and always led to unmanageable overhead and complicate the design of the hardware tasks for the system excessively thus negating part of the advantages of such a platform. The second decision was to allow for only pure HW VMs to be deployed. Since an FPGA SoC consists of a pair of ARM A9 CPUs and programmable logic, it is in principle possible to have VMs that are made up of a software part running on the ARM and a hardware part running on the programmable logic. The reason this was rejected for the PCP this is that the ARM A9 CPU currently available in FPGA SoC devices is not directly virtualizable itself, which increases the complexity of the task and reduces its appeal since the processors in the PCP already had their work cut out for them. Future FPGA SoCs like Xilinx's MPSoC [15] will include four ARM A53 CPUs which pack more punch and are directly virtualizable, which will make HW/SW VMs a more interesting endeavor.

8.3.1 Hardware

PCP allows for the deployment of tasks onto the programmable logic, which can be swapped in and out by the OpenStack controller as required. To accomplish this, we utilize a feature present in modern FPGA called dynamic partial reconfiguration (DPR). DPR allows part of the programmable fabric to be reprogrammed on the fly, while the remainder of the device continues to operate as intended. The PCP uses DPR to update the HW VM that is executed on the programmable logic on the fly. As such, the programmable logic in the programmable cloud platform is further subdivided into two parts, the static and the dynamic one, a typical arrangement in systems that utilize dynamic partial reconfiguration to change the functionality of a section of the device. This division is reflected in Fig. 8.1.

The static area is ancillary and enables the reconfiguration of the dynamic one, as well as the interconnection of the processor system (PS) with the dynamic area. It contains an AXI interconnect and AXI DMA module which connect the PS to the programmable logic (PL) as well as the PCAP that performs the reconfiguration of the dynamic area. The AXI DMA module contains two channels: one for data traffic to and from the HW VM and one used for sending monitoring data from the VM to the monitoring SW on the CPU. It doesn't implement Scatter–Gather DMA functionality as this would lead to higher resource usage due to the additional complexity and the

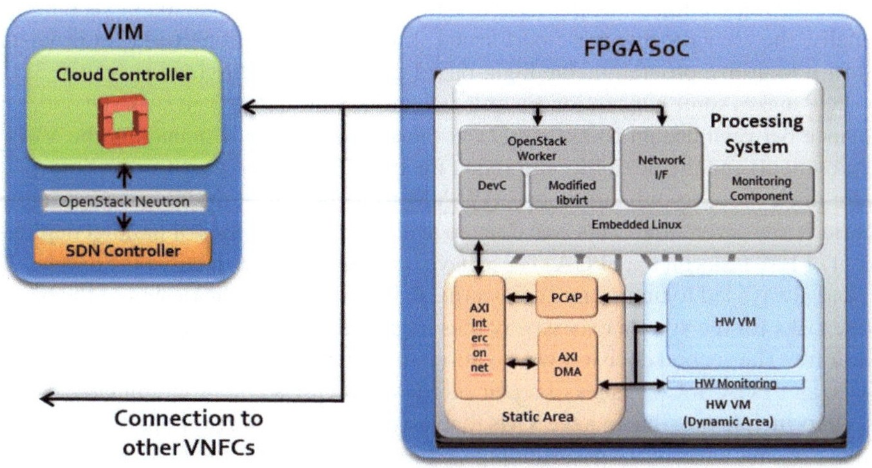

Fig. 8.1 Overview of the programmable cloud platform hardware/software architecture

expected benefit for our system is infinitesimal. The AXI communication fabric runs at 150MHz and has a width of 64 bits thus providing enough bandwidth to and from the dynamic area. An important goal for the static area is that it remains as small as possible to allow for the bulk of the programmable resources to remain available for the HW VMs. In the PCP the static area occupies approx. 6000 LUTs and 3500 FFs, which is less than 4.5% of the available resources on the used FPGA SoC device

The dynamic area is the larger part of the PL and is where the HW VM is deployed after receiving it over the OpenStack worker. The HW VM that is being deployed contains the user accelerator and a monitoring component as mentioned previously. This component is provided as part of the platform; however, the accelerator designer must feed it with the appropriate monitoring data which will then be send to the PS.

8.3.2 Software

A large part of the innovation in the programmable cloud platform comes from its software. At the core of the software is the modified, programmable logic-aware OpenStack (compute) worker node which is assisted by low-level software that interfaces with the Xilinx DMA driver, the DevC driver that feeds data into the PCAP, an FPGA component that is responsible for performing the actual DPR, and a pseudo libvirt library that poses as a counterpart to the worker and answers queries regarding the available resources of the device through an extended API as described in Sect. 8.3.3. As the FPGA SoC in the PCP is single-tenant, the libvirt is a, respectively, simple component. From a high-level perspective, workers executing a modified version of the compute service client are responsible for managing the VMs. Quite similar to its original purpose, workers can deploy, terminate, and reboot VMs, as well as run monitoring services to provide valuable information such as FPGA resource utilization. Given the simplicity of HW VM deployment in PCP, this module is at this stage, respectively, simple but it is bound to grow more complicated as the platform develops.

Additionally, the worker can take of the migration of an HW from one FPGA to another. This is a nontrivial exercise in an FPGA-based VM, as the entire configuration of the programmable logic has to be read out of the device, and then send to a remote node which will redeploy this to a different FPGA. In order for this to happen, execution of the HW VM has to halt first, which means that the worker stops accepting new data for this host, waits for all processing to complete, and then starts reading out the FPGA configuration, which is packaged and then passed on to the OpenStack controller over the network for deployment on a different node. The redeployment process is identical to the deployment of a new VM from a worker perspective. It is worth noting that since programmable logic configuration are completely bound to a specific device type, an HW VM that has been read out from a specific Zynq device (say the Zynq 7045 used in the PCP) can't be automatically deployed to a Zynq 7030 but only to a different Zynq 7045. This limits the flexibility of the migration process, but is a fundamental limitation of how configuration files for FPGAs work.

Furthermore, an overlay to the Xilinx DMA driver has been written which takes care of all the low-level driver functions necessary to create and initialize the DMA channels for reading and writing to the device and presents a character device driver in order to simplify data copying from the OpenStack worker to the PL. Special care has been given to make this a zero-copy driver to maximize the impact of the SW stack on performance.

An important aspect that must be taken care off in the software stack is the synchronization between data traffic and the deployment of a new VM. It is imperative that when the VM is retired and the programmable fabric is to be reconfigured, all pending transactions between the VM and the SW have been completed. This includes all DMA transactions for both the actual accelerator and any monitoring data that might be transferred at any given time. If stale data remains in the AXI interconnect, then this will lead to unpredictable behavior from which it might be impossible to recover. The SW stack takes special care of this by synchronizing the threads passing the data from the OpenStack agent on to the fabric. Thus when a new deployment request comes down for execution, the software is notified and stops receiving new transfers from that point on, while at the same time locking the deployment thread until all outstanding transactions have completed. The VM image (essentially the bitstream) can then be passed on to DevC driver which will perform the reconfiguration. The locks will then be released so traffic can start flowing again to and from the accelerator.

Finally, the software stack includes a monitoring component which reads the monitoring data of the dedicated AXI DMA channel, writes them into a local file, and transmits them to a remote server over the well-known curl utility, which has been cross compiled for the Zynq platform.

8.3.3 Making OpenStack FPGA-Aware

When it comes to deploying a VM using a multi-tenant OpenStack topology, the OpenStack compute worker node is only one side of the coin. The other side is the OpenStack controller node which is responsible for running services that direct and manage the deployment of tasks in the pool of worker nodes it supervises. It also provides a single point of access through API services for communication with the other components of OpenStack. Deploying an HW VM correctly requires not only an adapted worker but also and an FPGA-aware OpenStack Controller node. Related aspects of the controller's functionality that were taken into account and modified in order to make it FPGA-aware for the needs of the proposed solution are listed below:

- Adaptation of the scheduler service to recognize, locate, and manage programmable logic-based devices and assign HW VMs to them. The allocation of VMs to tenants is done based on several filtering functions that apply criteria to select the most suitable tenant that will host a VM. The selection is done based on a predefined set of metrics and information gathered from the monitoring services running on the worker nodes.

- The OpenStack Nova API service has been extended in order to include the calls to platform-specific functions. In this direction, additional commands have been added to the API that allows the management of the HW VM images by identifying and deploying them. So far the changes are command-line only. It goes without saying the extensions on the Nova API are backward compatible and are implemented with respect to the Nova developers policies. This is accomplished by taking advantage of the API Microversions framework that allows for specification of the API version that will be used in any circumstance.
- The Conductor service that also runs on the Controller node had to be slightly adapted in order to maintain access to the default relational database used by OpenStack for storing stateful information regarding the status of the platform. OpenStack's messaging queue is put to use to an extent that allows the brokerage of our system-specific messages between worker nodes, API service, scheduler, and the network service described below.

Another important aspect of deploying VMs over OpenStack is the provision and management of basic networking services. For the allocation of IP addresses and setup and configuration of the virtual networks of the host nodes (PCPs), a Nova-network service runs on the controller node that provisions services such as NAT and DHCP. A Nova-network client that executes on the worker node is responsible for configuring the network interface of the respective PCP-host. The Nova-network service was selected over the more advanced OpenStack Network Service (neutron) due to the added complexity and due to the fact that our requirements at this stage of the platform do not command for advanced networking topologies, or services such as load balancers and VPN which are offered by neutron.

Since the PCP is a compute only node, our modifications focused on the Nova component which is responsible for managing compute tasks but also extend to Glance, which administers the VM images used to deploy the VMs. Thus, the Glance component was modified to be able to manage HW VMs. Glance can now store these VMs and identify which ones are destined for FPGA SoCs and which not, avoiding mishaps during deployment.

It's worth noting that the changes performed are generic and do not limit PCP to being used with the development system currently used. Instead, any FPGA SoC can be used and even any FPGA provided its connected to a processor which can act as proxy for the OpenStack worker.

8.4 Experimental Results

PCP has been implemented using a Xilinx ZC706 development board which contains a Zynq xc7z045 FPGA SoC device. The board contains a Gigabit Ethernet interface which is used to connect the CPUs to the OpenStack controller and the traffic source and sink and an SD card which contains the boot image for the A9 CPUs. Thus, net-

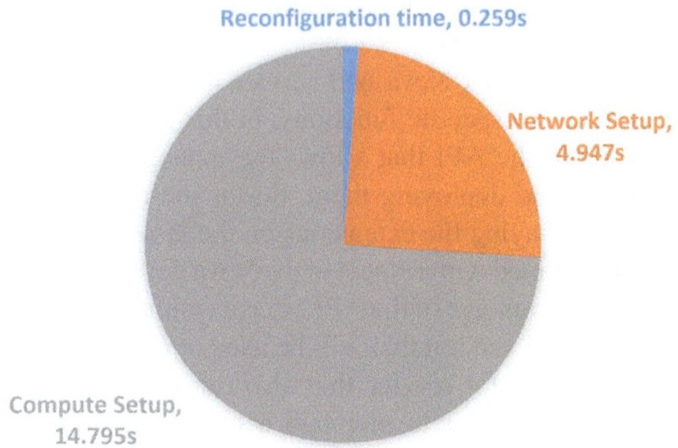

Fig. 8.2 Distribution of time lost during the deployment of an HW virtual machine

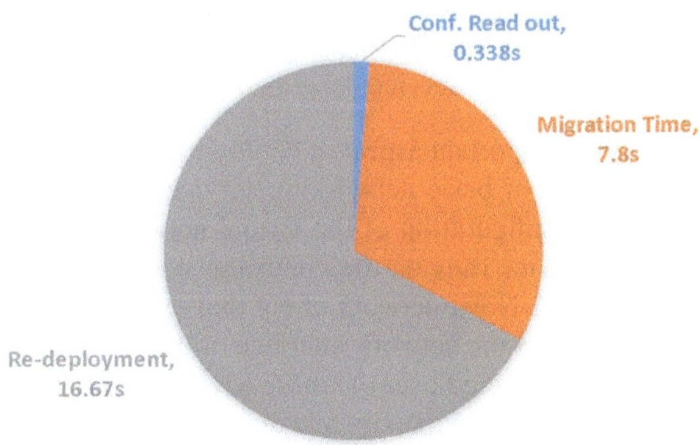

Fig. 8.3 Distribution of time required to migrate a virtual machine from one FPGA SoC node to another

work traffic entering the system is fed directly into the CPU for processing meaning all traffic to and from the accelerator goes through the A9s. This is mandatory since OpenStack has to perform network address translation on incoming data, which is performed at the SW level.

Evaluating the performance of our platform entails two separate aspects. First of all, we must ensure that the software stack running on the relatively weak A9 processors doesn't limit the capabilities of the programmable logic meaning that all OpenStack-related functionality is performed in a timely fashion. Furthermore, it is equally important to show that the HW VM running on the system's programmable fabric can indeed provide tangible benefits for a common application (Fig.8.5).

Another important aspect of VM deployment is migrating a VM from one node to another. This process was described in Sect. 8.3.2 and entails reading out the HW from the programmable logic via the DevC driver, sending it to the OpenStack controller and then redeploying it to a new device. We have measured the time the whole process takes and provided the data, breaking it down into its constituent parts (Fig. 8.2). Figure 8.3 illustrates the results averaged out over a set of 100 measurements.

The figure shows that reading the configuration out of the existing deployment (including stopping all traffic to that VM first) takes up only a small fraction of

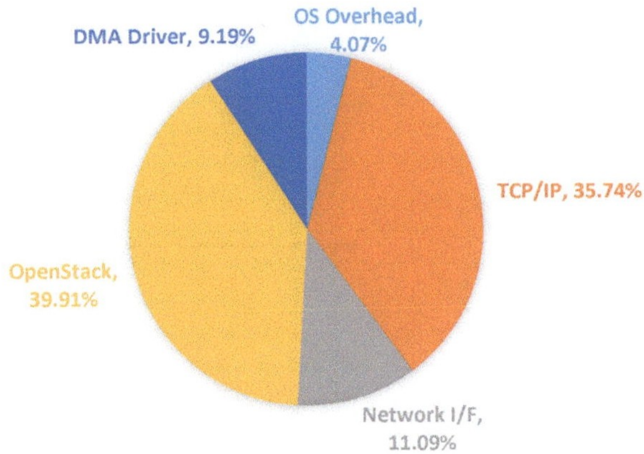

Fig. 8.4 Distribution of overhead when transmitting data through the SW stack

the total time it takes to migrate the VM completely. Migration time stands for the time required to send the data back to the OpenStack controller, while redeployment involves all the tasks detailed in Fig. 8.3, which are repeated when the VM is passed on to a new FPGA SoC node.

In order to evaluate the overhead caused by OpenStack processing on the Zynq's PS subsystem, we measure the time it takes to deploy one HW VM from the moment the appropriate command is passed to the controller, until the completion of the partial reconfiguration on the programmable fabric. Figure 8.3 illustrates how much time each part of the deployment cycle takes. The total process takes almost exactly 20 s with the bulk of the time being consumed by the setup of the compute instance followed closely by the creation of the necessary virtual network interfaces. While not negligible this time is on par with the deployment of SW VMs especially when one considers that an HW VM is readily available after being instantiated whereas an SW VM also has a substantial boot time. The actual reconfiguration of the fabric itself takes only a fraction of a second and can be safely ignored in this context.

The second facet that needs to be evaluated is whether the overhead incurred on the data when passing through the OpenStack worker during the HW VM's operation is low enough as well as the identification of sources of delay so that future improvements may be targeted accordingly. Figure 8.4 provides an overview of the form of a normalized pie chart of where in the software stack the time is spent when sending and receiving data traffic from and to the HW VM.

It's obvious from the above that the bulk of the processing cycles goes into TCP and OpenStack processing. TCP/IP is well known to be a processor cycle-hog [6], and this is verified on this platform as well even when only using a 1Gbps link. It is clear that the current software would not be able to support higher network data rates unless some sort of TCP acceleration (e.g., [12]) was utilized. This however would be implemented in the programmable logic and thus siphon resources away from the dynamic area and would still require the data to be shipped to the processor for the required OpenStack processing. OpenStack itself performs address translation on all incoming and outgoing packets, which is the reason behind its high cycle use.

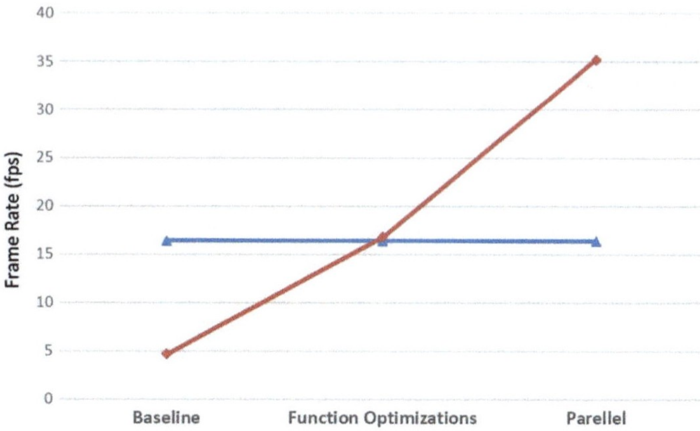

Fig. 8.5 Performance comparison of various implementations of the h264 decoder with a Xeon E5-2637 for a VGA video

A straightforward solution would be to simplify this processing given that PCP is a single-tenant platform, and thus, cohabitation of multiple user VMs is precluded. This is consigned to future work.

For the evaluation of the PCP, we have developed an h264 video decoder based on the OpenH264 software code provided by Cisco [5]. That code was reworked extensively to convert it to synthesizable C++ code which was then fed into Vivado HLS. There are three different versions of the design as shown in Fig. 8.5. The leftmost version represents the initial synthesizable version with no additional optimizations applied. The changes in this version can be essentially summed up in eliminating dynamic memory management prevalent in many areas of the original code. Thus the buffers and the context for each NAL [14] are now statically allocated.

The middle version includes optimizations made to the function hierarchy to minimize the impact of function calls, while the rightmost version represents a version that has been optimized and parallelized until the whole of the dynamic area was used up. It's worth noting that despite these modifications, the code was never rewritten from the ground up, and thus, it still maintains a structure which is not necessarily optimal for an HW module. Nevertheless, it shows that even with relatively modest reworking of existing SW code, considerable benefits can be reaped with the PCP.

Besides performance, resource use is an important parameter in FPGA designs. Table 8.1 provides the resource use numbers for the three variations of our h264 decoder. We can see that, as expected, the higher performing, more optimized variants consume significantly higher resources with the parallel implementation occupying a sizable chunk of the Zynq 7045 device used. Take in correlation with the performance number in Fig. 8.5 we can see that attaining a significant boost in comparison to standard x86 CPUs requires that a considerable part of the whole device is used (keep in mind that the number quoted on Table 8.1 does not include the overhead for the static area of our design, which is roughly 4.5% of the resources, as stipulated in

Table 8.1 h264 Decoder resource use on a Zynq 7045 FPGA SoC

	LUTs (%)	FFs (%)	BRAM (%)	DSP (%)
Baseline	31172/14.26	35151/8.04	19/3.57	36/3.96
Function optimization	71526/37.72	74062/16.94	36/6.67	78/8.7
Parallel	151381/69.25	148561/33.98	85/15.63	156/17.34
Total (Zynq 7045)	218600/100	437200/100	545/100	900/100

Sect. 8.3.2). This vindicates our decision to keep the dynamic area single-tenant in order to simplify the design of the system.

8.5 Conclusions

This paper introduced the programmable cloud platform, an FPGA SoC-based system which allows programmable logic to be exposed as a resource over a standard cloud environment like OpenStack. The system utilizes the devices ARM A9 processors to run the OpenStack agent and thus frees the programmable logic to accelerate tasks which are deployed as HW VMs. We use an application amenable to FPGA acceleration like h264 decoding to show that the platform allows for the swift deployment of the VNF and can feed it with sufficient data to reap tangible performance benefits. However, the cost of running OpenStack on a relatively weak processor is significant, and thus, one of the key issues to be tackled in the immediate future is the streamlining of the SW part to minimize this overhead. Additional work includes the extension of the platform's capabilities by integrating additional interfaces into the dynamic area.

Acknowledgements This work was undertaken under the EU FP7 Information Communications Technologies T-NOVA project, which is partially funded by the European Commission under grant 619520.

References

1. Asiatici M, George N, Vipin K, Fahmy SA, Ienne P (2016) Designing a virtual runtime for fpga accelerators in the cloud. In: 2016 26th international conference on field programmable logic and applications (FPL), pp 1–2
2. Bobda C, Majer A, Ahmadinia A, Haller T, Linarth A, Teich J (2005) The erlangen slot machine: increasing flexibility in FPGA-based reconfigurable platforms. In: Proceedings 2005 IEEE international conference on field-programmable technology, pp 37–42
3. Byma S, Steffan H, Bannazadeh G, Leon A, Chow P (2013) FPGAs in the cloud: booting virtualized hardware accelerators with openstack. In: 2014 IEEE 22nd annual international symposium on field-programmable custom computing Machines (FCCM), May 2013

4. Chen F, Shan Y, Zhang Y, Wang HF, Chang X (2014) Enabling FPGAs in the cloud. In: Proceedings of the 11th ACM conference on computing frontiers, May 2014
5. Cisco Inc. http://www.openh264.org
6. Foong AP, Huff TR, Hum HH, Patwardhan JR, Regnier GJ (2003) TCP performance re-visited. In: Proceedings of the 2003 IEEE international symposium on performance analysis of systems and software, ISPASS '03. IEEE Computer Society, Washington, DC, USA, pp 70–79
7. Ge X, Liu Y, Du DH, Zhang L, Guan H, Chen J, Zhao Y, Hu X (2014) OpenANFV: accelerating network function virtualization with a consolidated framework in openstack. In: Proceedings of the 2014 ACM conference on SIGCOMM, Aug 2014
8. Jouppi NP, Young C, Patil N, Patterson D, Agrawal G, Bajwa R, Bates S, Bhatia S, Boden N, Borchers A, Boyle R, luc Cantin P, Chao C, Clark C, Coriell J, Daley M, Dau M, Dean J, Gelb B, Ghaemmaghami T, Gottipati R, Gulland W, Hagmann R, Ho CR, Hogberg D, Hu J, Hundt R, Hurt D, Ibarz J, Jaffey A, Jaworski A, Kaplan A, Khaitan H, Koch A, Kumar N, Lacy S, Laudon J, Law J, Le D, Leary C, Liu Z, Lucke K, Lundin A, MacKean G, Maggiore A, Mahony M, Miller K, Nagarajan R, Narayanaswami R, Ni R, Nix K, Norrie T, Omernick M, Penukonda N, Phelps A, Ross J, Ross M, Salek A, Samadiani E, Severn C, Sizikov G, Snelham M, Souter J, Steinberg D, Swing A, Tan M, Thorson G, Tian B, Toma H, Tuttle E, Vasudevan V, Walter R, Wang W, Wilcox E, Yoon DH (2012) In-datacenter performance analysis of a tensor processing unit. In: To appear at the 44th International Symposium on Computer Architecture (ISCA), Toronto, Canada, June 24–28
9. Karras K, Manolakos ES (2008) An embedded dynamically self-reconfigurable master-slaves mpsoc architecture. In: 2008 international conference on field programmable logic and applications, pp 431–434
10. Pham KD, A Jain, Cui J, Fahmy S, Maskell DL (2013) Microkernel hypervisor for a hybrid arm-fpga platform. In: 2013 IEEE 24th international conference on application-specific systems, architectures and processors (ASAP), pp 219–226
11. Scott Sirowy AF (2008) Wheres the beef? why FPGAs are so fast. In: Microsoft Technical Report - MSR-TR-2008-130
12. Sidler D, Alonso G, Blott M, Karras K, Vissers K, Carley R (2015) Scalable 10 gbps TCP/IP stack architecture for reconfigurable hardware. In: 2015 IEEE 23rd annual international symposium on field-programmable custom computing machines (FCCM), pp 36–43
13. Wang K, Angstadt K, Bo C, Brunelle N, Sadredini E, Tracy T, Wadden J, Stan M, Skadron K (2016) An overview of micron's automata processor. In: 2016 international conference on hardware/software codesign and system synthesis (CODES + ISSS), pp 1–3
14. Wiegand T, Sullivan GJ, Bjontegaard G, Luthra A (2003) Overview of the h.264/avc video coding standard. IEEE Trans Circ Sys Video Technol 13(7):560–576. http://www.rsc.org/dose/title-of-subordinate-document Cited 15 Jan 1999
15. Xilinx Ultrascale+ MPSoC. https://www.xilinx.com/products/silicon-devices/soc/zynq-ultrascale-mpsoc.html

Chapter 9
Energy-Efficient Servers and Cloud

Huanhuan Xiong, Christos Filelis-Papadopoulos, Dapeng Dong, Gabriel G. Castañé, Stefan Meyer and John P. Morrison

9.1 Introduction

As the sizes of cloud infrastructures continue to grow, the complexity of the cloud is becoming more and more difficult to manage. Currently, centralised management schemes dominate and there are already signs that these are no longer fit for purpose. Elasticity, for example, (the ability of the cloud to respond to rapidly changing demands for resources) is currently being supported by over-provisioning. Over-provisioning is a strategy of effectively underutilising hardware so that some is always available to absorb unpredictable peaks in demand. This strategy is not sustainable, since the infrastructure costs and the energy it consumes, even when idle, are significant. In 2010, Gartner Research [14] reported that the average server utilisation in large data centres is 18%, while the utilisation of x86 servers is even lower at 12%. These results confirmed earlier estimations that the average server utilisation is in the range of 10–30% [7]. Subsequent studies have not contradicted these findings [16, 19].

Cloud computing is evolving from its homogeneous roots and is being seriously regarded by once highly specialised application domains like high-performance computing (HPC).

To support this trend, heterogeneity is a must [11]. HPC technology trends in coprocessors are on the increase and NVIDIA GPGPUs and Intel Xeon Phi are gaining increased traction. Low-power processors are beginning to find their place in the HPC ecosystem and HPC public cloud revenue could range from $1.56 billion (low forecast) to $3.7 billion (high forecast) by 2017 and for HPC public custom

H. Xiong · D. Dong · G. G. Castañé · S. Meyer · J. P. Morrison (✉)
Department of Computer Science, University College Cork, Cork, Ireland
e-mail: j.morrison@cs.ucc.ie

C. Filelis-Papadopoulos
Department of Electrical and Computer Engineering, Democritus University of Thrace, University Campus, Kimmeria, GR 67100 Xanthi, Greece

C. Kachris et al. (eds.), *Hardware Accelerators in Data Centers*,
https://doi.org/10.1007/978-3-319-92792-3_9

cloud computing, worldwide revenue could range from $0.87 billion (low forecast) to $1.5 billion (high forecast) in the same period.

Incorporating energy-efficient heterogeneous resources help to address power consumption in the cloud. However, their inclusion also adds to the complexity of the already overburdened cloud management scheme and this must be explicitly addressed. Energy efficiency in cloud computing can be considered as a complex optimisation problem, which attempts to minimise power consumption while satisfying quality-of-service (QoS) or service-level agreement (SLA) requirements specified by users. Energy-aware resource provisioning and allocation is a way to improve energy efficiency without violating the negotiated SLAs or application performance [3, 4, 15]. Research shows that the objective of energy efficiency is not an independent or stand-alone issue from other cloud resource management objectives, such as QoS/SLA, resource utilisation and workload performance (e.g. execution time, intensity), for example, and, unfortunately, there would appear to be no single equation capable of expressing all the inter-dependences between the multiple objectives.

The CloudLightning project takes a novel route, making use of self-organisation techniques to address the problems emerging from the confluence of issues in the emerging cloud: rising complexity and energy costs, problems of management and efficiency of use, the need to efficiently deploy services to a growing community of non-specialist users and the need to facilitate solutions based on heterogeneous components. Thus, this approach attempts to address

- Energy efficiency.
  Self-organisation is a powerful tool for addressing complexity of large- to hyper-scale cloud resource management. It has proven itself time and time again in nature and has been applied successfully in complex engineering projects [12]. Of self-organisation, Alan Turing once observed that global order arises from local interactions. We contend that when self-organisation is applied to self-management, local interactions can give rise to scalable global management. Moreover, given the appropriate evolutionary stimuli, the resultant global management can be optimised for specific characteristics. CloudLightning proposes a self-organised self-managed (SOSM) framework for providing energy-efficient cloud resource provisioning and allocation (see Sect. 9.3).
- Improved accessibility to cloud.
  The CloudLightning SOSM system provides cloud service consumers with a user-friendly service-level interface to explicitly declare their requirements for service delivery. Through the assembly of dynamic resource coalitions, the SOSM system automatically and intelligently locates the required resources and chooses the most appropriate configuration to deliver that service, while respecting both the user-level SLA and the business objectives of the cloud service providers (CSPs). CSPs are thus enabled to provide energy-efficient, scalable management of their cloud infrastructures and better overall utilisation of service.
- Supporting heterogeneity.

To support heterogeneity, CloudLightning brings various coprocessors into existing homogeneous cloud environments. These include graphic processing units—GPUs, many integrated cores—MICs, field programmable gate arrays—FPGAs. The availability of different resource types can alter the way that solutions are designed. Mapping a problem onto an architecture specifically designed with that problem in mind can greatly improve efficiency and simplify implementation. Secondary benefits are often obtained such as speed, improved precision of solution and reduced power consumption. These are important drivers for both the cloud provider and for the end user. CloudLightning also provides a plug-in mechanism for incorporating heterogeneous resources into the self-organising cloud. Pertinent characteristics of these resources are surfaced, through the plug-in mechanism, to the end user via the service description language; making these resources easier to consume.

The three objectives listed above are tightly coupled aspects of the CloudLightning system. A complete description of the CloudLightning system is necessary to express the subtle interplay between the objectives and the architecture of the solution needed to address them. However, here the CloudLightning system is described predominantly from the energy efficiency perspective and the advantages that flow from exploiting hardware accelerators and the challenges associated with balancing energy consumption with improved service delivery.

The remainder of this chapter is organised as follows. Section 9.2 presents the CloudLightning hierarchical architecture and its main components. The self-organised self-managed framework with respect to energy-efficient resource management is described in Sect. 9.3. Finally, Sect. 9.4 presents the evaluation of our proposed approach and Sect. 9.5 concludes with some final thoughts.

9.2 CloudLightning Architecture

Large-scale data centres typically make use of a hierarchical model for organising the compute, storage and network infrastructures. The warehouse scale computer (WSC) [2] is a typical hierarchical architecture widely used by companies like Google, Yahoo, Amazon, Facebook, Microsoft and Apple [1, 9, 17] for this purpose.

The CloudLightning architecture is also a hierarchical organisation of physical infrastructure but unlike traditional organisations it makes use of a resource management framework that is logically hierarchical. The bottom layer of this framework hierarchy consists of many resource managers. These managers are autonomous and, in contrast to traditional systems, each manages a relatively small number of physical resources. Since the number of physical resources is restricted, each manager can efficiently control the collection- allocating tasks and, where appropriate, virtualising resources in response to service requests. This arrangement is self-limiting in the sense that an increase in the number of physical resources attached to such a manager spontaneously results in a new manager being brought into existence to

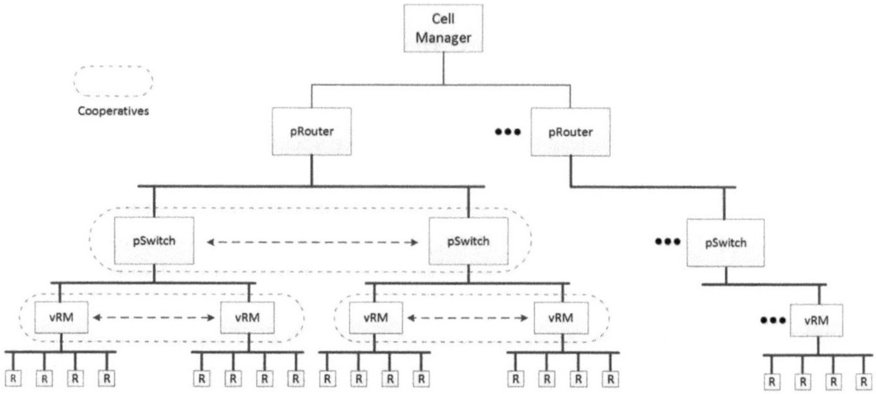

Fig. 9.1 A representation of the CloudLightning architecture

assume the control over the extra resources should the increase in physical resources exceed a specific cost management threshold. In this way, a first step is taken to tackle the problem of scalability in resource allocation. However, this enhancement to the topology of the hierarchy forms only a partial solution, since no mechanism is yet provided to identifying an appropriate resource manager, from many potential candidates, that will make the final resource allocation decision.

The CloudLightning architecture is depicted in Fig. 9.1.

9.2.1 Cell

At the top of the hierarchy, a cell represents the entire set of physical resources. These are partitioned into different hardware types (including CPUs, CPU–GPU pairs, CPU–MIC pairs and CPU–FPGA pairs) and each partition is accessed via a dedicated pRouter.

9.2.2 pRouter

Each pRouter provides access to hardware resources of the same type which, in turn, is managed by a specific resource abstraction method (such as OpenStack Nova[1] to manage virtual machines on commodity servers, Kubernetes,[2] Mesos[3] [10] and/or

[1]OpenStack Nova: http://docs.openstack.org/developer/nova/.

[2]Kubernetes: http://kubernetes.io/.

[3]Apache Mesos: http://mesos.apache.org/.

Docker Swarm[4] to manage containers on GPUs and MICs, and OpenStack Ironic[5] to manage bare metal servers using DFEs).

Some examples of the constitution of the pRouters in CloudLightning architecture include

- OpenStack-CPUs-VMs: Commodity machines (CPUs) pre-installed with OpenStack services.
- OpenStack-FPGAs-accelerators: FPGAs configured as compute accelerators using the OpenStack Nova service.
- Mesos-GPUs-accelerators: GPUs configured as compute accelerators using the Mesos framework and Docker Engine.
- Marathon-MICs-accelerators: Marathon as the resource manager managing a cluster of Xeon servers with attached MICs.

Each pRouter is connected to one or more pSwitches.

9.2.3 *pSwitch*

pSwitches are used to further partition the resource space into smaller and more manageable domains composed of multiple virtual Rack Managers (vRMs). However, the number of pSwitches per pRouter is not fixed and can change over time. Similarly, the number of vRMs, being managed by each pSwitch, can also change over time in response to dynamic growth and shrinkage of the resource fabric.

pSwitches and vRMs can self-organise by exchanging constituent members. Thus, two of more, pSwitches may exchange control over a subset of their respective vRMs and, similarly, two within groups, which will be called cooperatives, to emphasise their self-organising nature. To prohibit the creation of cooperatives with different resource types, pSwitch cooperatives cannot span pRouters. Similarly, to minimise administrative overhead, vRM cooperatives cannot span pSwitches.

As the CloudLightning system evolves, it is anticipated that the number of pSwitches connected to a pRouter will change and will converge to some optimal number with respect to some global goal set by the cloud service provider. This goal is expressed as a vector of weights that are propagated down through the management hierarchy and alter the perceived importance of the underlying behaviours. As part of the self-organisation process, pSwitches and vRMs can be created, destroyed, merged and split.

---

[4]Docker Swarm: https://github.com/docker/swarm/.

[5]OpenStack Ironic: http://docs.openstack.org/developer/ironic/deploy/user-guide.html.

9.2.4 *vRack Manager*

At the bottom of the hierarchy are a collection of resource managers known as the virtual Rack Managers (vRMs). These are responsible for the efficient management of the collection of resources directly under their control. vRMs also communicate weighted status information pertaining to these resources upwards through the hierarchy.

It is anticipated that the number of vRMs connected to a pSwitch and the number of pSwitches connected to a pRouter will change and will converge to some optimal number; derived from the weights coming from the pRouter and from the vRM's and pSwitch's efforts to converge to a local goal state. As part of the self-organisation process, vRMs and pSwitches can be created, destroyed, merged and split. Furthermore, they may exchange control over resources in an effort to maximise resource utilisation, to minimise energy consumption and to optimise management utility.

9.3 Hyper-scale Resource Management for Energy Efficiency

In the CloudLightning system, the process of identifying an appropriate resource manager to affect the next resource allocation decision is distributed throughout the entire logical hierarchy. It begins at the vRM level, where information relating to the functional capabilities, and the non-functional behaviours, of its constituent resources forms a view of these resources, which is then propagated upwards through the hierarchy. In this upwards propagation, and at each intermediate level in the hierarchy, this information may be combined into a higher level view, in many different ways, with similar information emerging from different elements from the lower level of the hierarchy. These views are called *Perceptions* and are used to guide the resource allocation requests entering the system at the top of the hierarchy. The contention is that the most appropriate resource allocation is to be found by following the path exhibiting the greatest perception, since this path simultaneously maximises the chances of locating the requisite resources and of optimising the non-functional behaviours.

CloudLightning develops a strategic self-organised self-managed framework to support distributed resource allocation decisions and that can be dynamically populated with strategies to reflect the ever-growing number of diverse objectives as they become evident in the evolving cloud infrastructure. In the CloudLightning approach, cloud service providers can define various strategies by which cloud resources are allocated, and system/cloud objectives (from energy efficiency perspective) are attained.

In the CloudLightning approach, these objectives can be expressed as assessment functions (see Sect. 9.3.1). These outputs of assessment functions are aggregated as they pass upwards through the hierarchy, the components in each layer of the hierar-

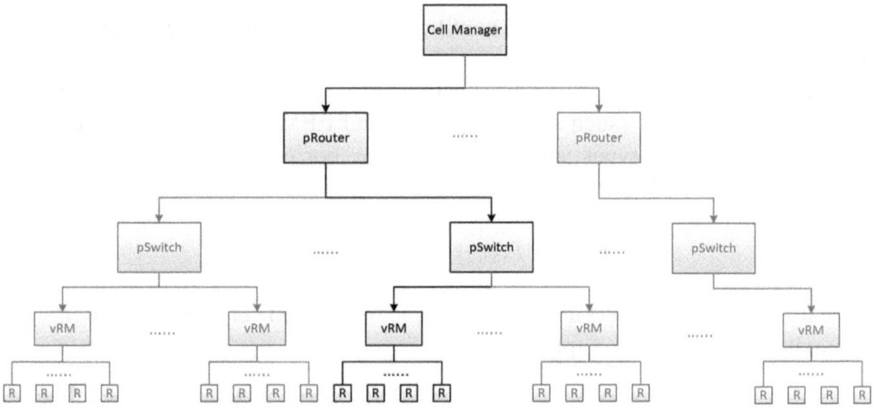

Fig. 9.2 An example of resource allocation path based on SI

chy (i.e. cell manager, pRouters, pSwitches and vRMs layers), and then use various strategies to calculate a value known as the suitability index (SI) (see Sect. 9.3.2). This index indicates the suitability of the underlying region to host the next service request. Thus, a pathway guiding a service request to the place where it is most likely to find resources to host is determined by always choosing to enter a region having the greatest SI. This is illustrated in Fig. 9.2.

9.3.1 Assessment Functions

In CloudLightning, assessment functions output metrics are used for monitoring and reflecting the state of the cloud infrastructure, including functional and non-functional behaviours such as computation performance, power consumption, and management cost. Achieving energy efficiency in the cloud is not simply a matter of reducing power consumption in isolation. Power reduction must be done in the context of guaranteeing workload computation performance; maximising the computation performance within a particular power consumption budget. Therefore, it makes sense to express the various aspects being captured by the CloudLightning assessment functions in terms of energy efficiency.

Performance per watt is a measure of the energy efficiency of a particular computer architecture or computer hardware.

$$f_{hw} = \frac{Performance}{Power}. \tag{9.1}$$

Computational performance could be evaluated in measurable, technical terms, using one or more metrics, such as CPU/memory utilisation, throughput, floating point operations per second (FLOPS), millions of instruction per second (MIPS) and bandwidth.

Usually, the performance of a hardware configuration can be measured by any appropriate benchmark (such as SPECpower[6] and EEMBC.[7]) Power models vary for different hardware types and hardware usage. For example, a commonly used linear power model [4] for CPU-based servers is

$$P(u) = P_{\min} + (P_{\max} - P_{\min})u, \tag{9.2}$$

where $u \in [0, 1]$ is the CPU utilisation, $P_{\min}$ is the idle power consumption, $P_{\max}$ is the maximum power consumption.

9.3.2 *Suitability Index*

In CloudLightning, the concept of **Suitability Index (SI)** is created for measuring how close a component is to its desired state, and hence how suitable its operating characteristics are for contributing to the global goal.

$$\underset{\mathbf{w}^{\ell},\mathbf{m}^{\ell} \in \mathbb{R}^N}{\arg\max}\ \eta\left(\mathbf{w}^{\ell}, \mathbf{m}^{\ell}\right), \tag{9.3}$$

where $\mathbf{w}^{\ell}$ is an N-dimensional vector of weights corresponding to the impetus in the ℓ-th level and $\mathbf{m}^{\ell}$ is an N-dimensional vector of metrics corresponding to the perception in the ℓ-th level. Generally speaking, $\mathbf{w}^{\ell}$ presents the influence factor from the upper level indicating the perspectives from application characteristics, system objectives, service-level agreement, etc., and $\mathbf{m}^{\ell}$ is the perception value (i.e. mean and/or maximum) of the lower level giving the average and/or the best performance view over the underlying system.

Overall, in terms of different characteristics of assessment functions, SI can be used to indicate the most suitable location to host incoming service requests. Thus, when the assessment functions are chosen to reflect the energy consumption with respect to different aspects of computation performance, the SI will indicate the most energy-efficient location in the cloud resource fabric for hosting the next incoming service request. Similarly, if the assessment functions are chosen to reflect the management costs associated with different cloud configurations, the SI will indicate to the most efficient place to host the next incoming service request, with respect to that management cost.

Figure 9.3 depicts the various applications of using the suitability index to achieve different goals.

---

[6]SPECpower: https://www.spec.org/power_ssj2008/.

[7]EEMBC: http://www.eembc.org/.

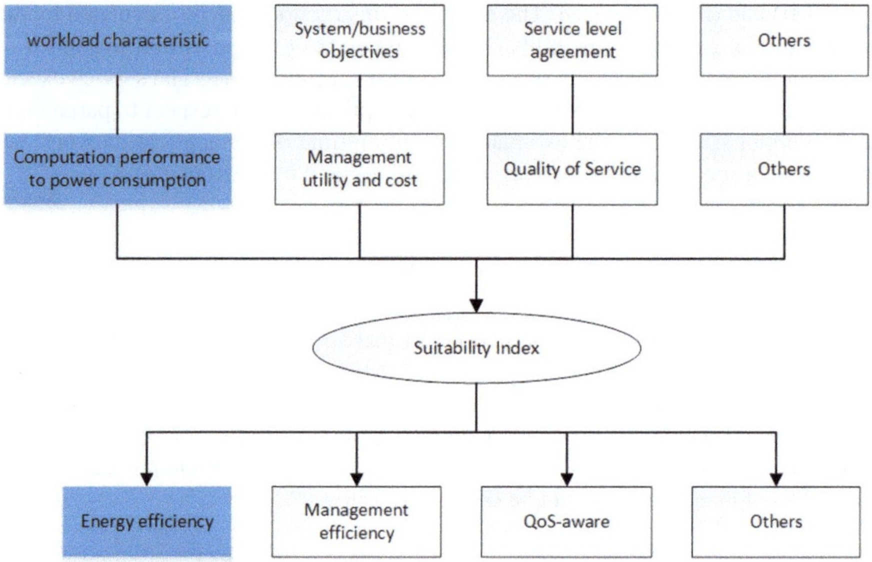

Fig. 9.3 Suitability index use cases

9.3.2.1 SI Strategy on CM

In the CloudLightning system, the cell manager needs to decide on which resource type is the most appropriate for hosting the next incoming service request. When user's SLA objectives are satisfied, a choice can be made from the remaining types, which maximises system objective (e.g. maximising energy efficiency).

The energy consumption for a specific service is often modelled as the integral of the power consumption function over the execution time for completing the application (mostly for batch and HPC workloads, not for the long-run web applications)

$$E = \int_{t_0}^{t_1} P(u(t))dt, \tag{9.4}$$

where $u(t)$is the CPU utilisation function of time, which may change over time due to the workload variability. $t_1 - t_0$ denotes the application execution time.

The SI strategy on cell manager is to find the most suitable pRouter (that is, a specific resource type) for a specific service workload with the respect to energy efficiency. Therefore, the cell manager has to calculate/predict the energy consumption for each possible hardware type which might run that application workload based on their current state and with knowledge of the service characteristics and hardware characteristics.

In some cloud simulators [5, 13, 18], service workloads can be modelled with three parameters: (i) input size, (ii) processing length, measured in millions of instructions (MI) and (iii) output size. The number of instructions can be calculated by the computational capacity (capability) of the processors (can be virtual machines or accelerators) multiplied by the service workload execution time. The service execution time can be computed from profiling the application with respect to parameters such as input size. Thus, the estimated execution time of the application workload running on a specified hardware resource (represented by a pRouter i) is

$$t = \frac{N^{MI}}{N_i^{MIPS}}, \tag{9.5}$$

where N^{MI} is the number of instructions in that service, and $N_i^{MIPS}(MAX)$ is the maximum number of instructions per second (MIPS) from pRouter i.

The total computational capacity of pRouter i hosting $N_{servers}$ number of servers, N_{acc} number of accelerators per server, with C_{proc} the combined computational capacity (in MIPS) of all processors (CPUs) of a server and C_{acc} the computational capacity (in MIPS) of an accelerator, can be defined as follows:

$$C_i^{pr} = C_{proc} N_{servers} + C_{acc} N_{acc} N_{servers} \ (MIPS). \tag{9.6}$$

The computational capacity (in MIPS) of the i-th pRouter (C_i^{pr}) can be defined in terms of servers executing a task

$$C_i^{pr} = C_{proc} \frac{N_{proc}^u}{N_{proc}} + C_{acc} N_{acc}^u \ (MIPS), \tag{9.7}$$

where N_{proc} is the number of processing units per server, $N_{proc}^u \in [0, N_{proc} N_{servers}]$ is the number of utilised processing units with respect to all servers under a pRouter and $N_{acc}^u \in [0, N_{acc} N_{servers}]$ is the number of utilised accelerators with respect to all servers under a pRouter. The result of Eq. (9.7) would be similar for all resources at the beginning since no task has entered the system. Thus, in order to distinguish between resources, a random number can be added on the SI.

In order to compute the power consumption per pRouter we can utilise the model of Eq. (9.2). The power consumption for the accelerators is considered to be binary, since accelerators are either used or not and cannot be shared between virtual machines. The power consumption of the pRouter i can be estimated as follows:

$$P_i^{pr} = \left(P_{\min} + P_{\min}^{acc} N_{acc}\right) N_{servers} + \left(P_{\max} - P_{\min}\right) \frac{N_{proc}^u}{N_{proc}} + \left(P_{\max}^{acc} - P_{\min}^{acc}\right) N_{acc}^u \ (W), \tag{9.8}$$

where N_{proc} is the number of processing units per server, $N_{proc}^u \in [0, N_{proc} N_{servers}]$ is the number of utilised processing units with respect to all servers under a pRouter and $N_{acc}^u \in [0, N_{acc} N_{servers}]$ is the number of utilised accelerators with respect to all servers under a pRouter. The quantities $P_{\min}^{acc}$ and $P_{\max}^{acc}$ are the idle and the maximum power consumptions of an accelerator, respectively.

Therefore, the computation of the suitability index (SI) of each pRouter, at the cell manager, can be performed as follows:

$$SI_i^{pr} = C_i^{pr}/P_i^{pr}(MIPS/W). \tag{9.9}$$

This metric can be used in conjunction with the available (status) information to guide a task to the most efficient resource in terms of higher $MIPS/W$. The aforementioned equations can be reformed to take into account the overcommitment of resources. Guiding the task below the pRouter level can be performed using the same definition for the SI or the definition given in [8].

Similarly, the computation of the SI of each pSwtich (at a pRouter), and the computation of the SI of each vRM (at a pSwitch), will follow the same pattern as described in Eqs. (9.7), (9.8) and (9.9).

9.3.2.2 SI Strategy on vRM

The SI strategy on vRMs is quite different from the cell manager, pRouters and pSwitches, since the vRMs have up-to-date state information for all the resources under their control. vRMs can make accurate and timely decisions about the resource allocations with the respect to energy efficiency locally.

In CloudLightning, three strategies are applied to achieve energy-efficient resource allocation.

(1) Best energy-efficient node

The simplest strategy for a vRM to achieve energy efficiency is to deploy the virtual machine (VM) onto the most power efficient node (i.e. server). This strategy would appear to prioritise server utilisation, however, success in selection depends on there being sufficient resources available from that server to satisfy the VM's requirement.

(2) Bin-packing

The objective of this approach is to minimise the energy consumption by placing VMs onto the minimum number of hosts. The model can be described as follows [6]:

$$\begin{aligned}
\min \sum_{i\in V}\sum_{j\in H} p_{ij}v_{ij} + \sum_{j\in H} P_j h_j & & & \\
\text{s.t.} \sum_{i\in V} r_i v_{ij} &\leqslant Ch_j & & \forall j \in H \\
\sum_{j\in H} v_{ij} &= 1 & & \forall i \in V \\
v_{ij} &\leqslant h_j & & \forall i \in V, j \in H \\
v_{ij} &\in \{0, 1\} & & \forall i \in V, j \in H \\
h_j &\in \{0, 1\} & & \forall j \in H,
\end{aligned} \tag{9.10}$$

where H is the set of hosts, V is the set of VMs in the cloud fabric, the objective is to decide how to rearrange V on H such that the total power consumption in the system is minimised. All $v \in V$ requirements r_i, such as CPU, memory and storage, must be satisfied by the targeting host; each h has a resource capacity limit C. The total power consumption is the sum of power p_{ij} consumed by CPUs of each VM i on host j, plus a fixed power P_j consumed by the other components of host j, such as memory and I/O. Let $h_j = 1$ represent choosing host j to be switched on, and 0 otherwise. Also, let $v_{ij} = 1$ represent the assignment of VM i to host j, and 0 otherwise. The first constraint enforces the capacity limit on each host. The second constraint ensures that each VM is assigned to exactly one host. The third constraint guarantees a host to be switched on if and only if a VM has been assigned to that host. The last two constraints indicate the state of a VM or host is either to be on or off.

However, this approach does not take the service workload characteristics and overcommitment into account, the next strategy applies the bin-packing with overcommitment ratio to further optimise the energy efficiency within a vRM.

(3) Bin-packing with overcommitment

There are three basic types of scheduling execution for VMs residing on a server: space-sharing, time-sharing and hybrid (time–space) policies.

- Space-sharing policies divide the system into partitions of processors so that more than one task can run on the system simultaneously; each on its own group of processors.
- In a time-sharing model, computer resources are committed and tasks are executed at the same time among many users. Time-sharing policies adopt pre-emption to alternate processors among a number of tasks that is usually determined by the multiprogramming level.
- Hybrid models (space–time sharing) combines the benefits of the two aforementioned types of policies. In a hybrid model, the system is divided into multiple processor groups and each group adopts a time shared policy by a distinct set of application tasks.

The time-sharing and hybrid model are commonly used in overcommitment of CPU and memory resources, which can increase server utilisation. However, for compute-intensive workloads, overcommitment of resources can reduce performance and subsequently increase energy consumption, since the tasks are competing for resources. While for the communication-intensive workloads, the overcommitment of CPU and memory resources can be less harmful with respect to performance.

Thus, each vRM can customise its own overcommitment ratio with respect to the available physical resources (i.e. CPU, memory and network) and the workload characteristics, dynamically adjusting the overcommitment ratio of its associated resources to maximise the energy efficiency without violating the service performance.

Table 9.1 Characteristics of the resources

Resource	C_{proc}	C_{acc}	N_{proc}	N_{acc}	$N_{servers}$	$P_{\min}$	$P_{\max}$	$P^{acc}_{\min}$	$P^{acc}_{\max}$
1	160,000	0	16	0	500	100	500	0	0
2	160,000	480,000	16	4	500	100	500	50	250

9.4 Evaluation

In this section, the scheme for computing the SI, in the cell manager level, is evaluated. Without loss of generality a cell is considered as having two types of hardware. The characteristics of these resources are given in Table 9.1. For simplicity, tasks are considered to fully utilise underlying resources and implementations exist for both tasks. The length of the task queued at each time step was computed as a rounded random value, obtained the uniform random distribution, in the interval $[0, 0.85]$. Thus, the total number of tasks that entered the system was 35,629. The tasks required $2, 4, 8, 16$ virtual cores and $1, 2, 3, 4$ accelerators, respectively, following uniform random distribution. The number of instructions for a task was computed in the interval $[100, 5000]$ *MI* using a uniform random distribution. The simulated time was 172,800 s. Tasks enter the system with respect to the aforementioned parameters for the first 86,400 s, while for the last 86,400 the system is left to finish execution without incoming tasks. The energy consumption is computed as the integral of power consumption over time. The integral is computed numerically using the rectangle method.

The first experiment concerns the computation of the SI using Eqs. (9.6), (9.8) and (9.9). In Fig. 9.4, the suitability indices (*MIPS/W*) computed with Eqs. (9.6), (9.8) and (9.9) for the two types of hardware are given. In the beginning, and up to time step 1841, tasks flow to the hardware of Type 2, since its SI has the highest value. When the value of the SI of hardware Type 2 reaches the value of the SI of hardware Type 1, tasks start to flow to the resources of Type 1. From time step 1842 to time step 2350, the two pRouters are competing for the incoming tasks, since the SIs are almost similar in value. From the time step 2351 to time step 86,000, the pRouter hosting hardware of Type 2 is almost completely utilised; thus, most of the tasks flow to the pRouter hosting hardware of Type 1. This process leads to an overall power consumption of 25.5129 MWh.

The second experiment concerns the computation of the SI using Eqs. (9.7), (9.8) and (9.9). In Fig. 9.5, the suitability indices (*MIPS/W*) computed with Eqs. (9.7), (9.8) and (9.9) for the two types of hardware are given (first task arriving to pRouter hosting hardware of Type 1). In this experiment the initial value of both SIs is 0, thus the first task flows to the first available pRouter (e.g. Type 1). The following tasks also flow to pRouter hosting hardware of Type 1 until it is almost completely utilised. Then, tasks start to flow to the pRouter hosting hardware of Type 2. Due to the large value of the SI of the pRouter hosting hardware of Type 2 the majority of the tasks continue to flow there, while a small fraction of the tasks flow to the pRouter hosting hardware of Type 1. This process leads to an overall power consumption of 26.2064 MWh.

The initial hardware choice dictates the first task flows and impacts the direction taken by the following tasks. In Fig. 9.6, the suitability indices (*MIPS*/*W*) computed with Eqs. (9.7), (9.8) and (9.9) for the two types of hardware are given (first task arriving to pRouter hosting hardware of Type 2). This process leads to an overall power consumption of 25.6172 MWh. Choosing the fastest hardware type first leads to improved energy consumption, since the CPU–Accelerator pair executes tasks faster.

The energy consumption using Eqs. (9.6), (9.8) and (9.9) is slightly reduced compared to the other two approaches.

The last experiment concerns the computation of the SI following the initial approach given in [8], with

$$f_1(N^u_{proc}) = C\frac{N^{total}_{proc} - N^u_{proc}}{N^{total}_{proc}}, \tag{9.11}$$

$$f_2(N^u_{proc}) = \frac{P_i(N^{total}_{proc} - N^u_{proc})}{PN^u_{proc} + P_i(N^{total}_{proc} - N^u_{proc})}, \tag{9.12}$$

and the vector of weights $\mathbf{w} = [1\ 1]$. The suitability index is computed as $SI = w_1 f_1 + w_2 f_2$, [8]. For the two types of hardware (1 and 2), the values of the relative computational capability C were 1.0 and 3.0, respectively [8]. The values for the

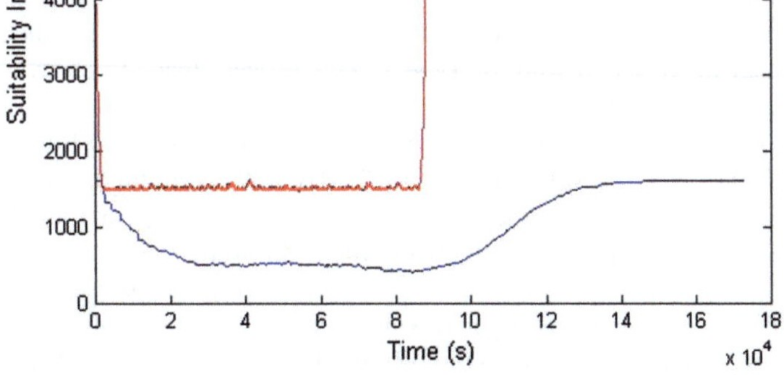

Fig. 9.4 Suitability indices (*MIPS*/*W*) computed with Eqs. (9.6), (9.8) and (9.9) for the two types of hardware

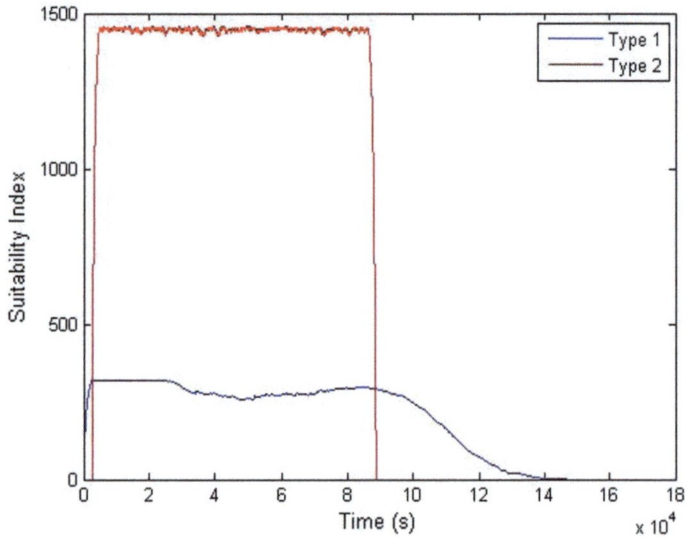

Fig. 9.5 Suitability indices (*MIPS/W*) computed with Eqs. (9.7), (9.8) and (9.9) for the two types of hardware (first task arriving to pRouter hosting hardware of Type 1)

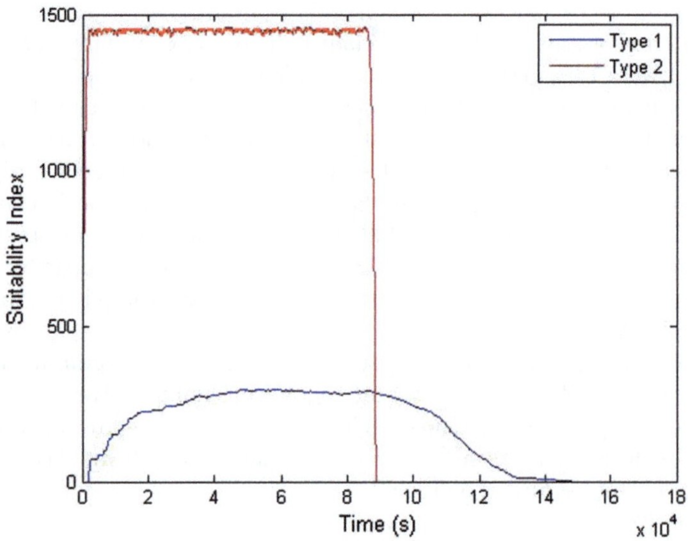

Fig. 9.6 Suitability indices (*MIPS/W*) computed with Eqs. (9.7), (9.8) and (9.9) for the two types of hardware (first task arriving to pRouter hosting hardware of Type 2)

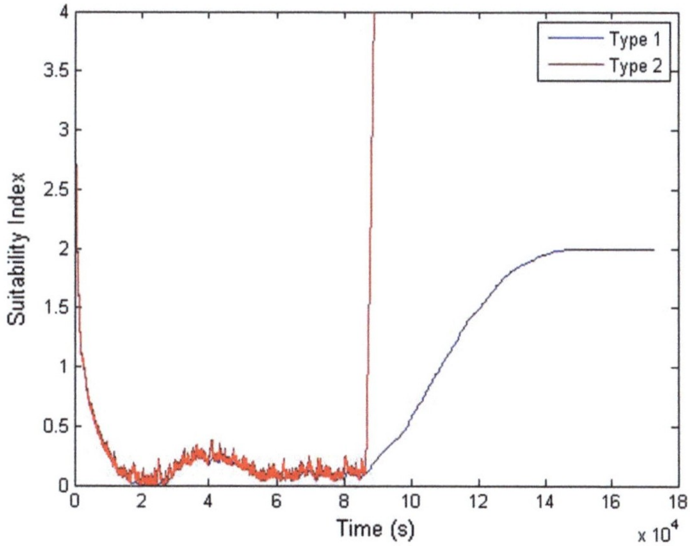

Fig. 9.7 Suitability indices computed with respect to the original design given in [8]

relative power consumption P and the relative idle power consumption P_i for hardware Type 1 were 1.0 and 0.2, respectively, while for hardware Type 2 were 1.5 and 0.2, [8]. For hardware Type 2, Eqs. (9.11) and (9.12) are computed with respect to the total number of accelerators as well as the number of utilised accelerators. In Fig. 9.7, the suitability indices computed with respect to the original design given in [8] are presented. In this approach, the two pRouters continuously compete for acquiring tasks from time step 0–86,400. The pRouter hosting hardware of Type 2 is the first to receive tasks until the point where its SI becomes comparable to the SI of the pRouter hosting hardware of Type 1. The energy consumption of the system was 27.1687 MWh.

The proposed approaches for computing the SI lead to an overall improvement between approximately 3.54–6.1% for the energy consumption. The computation of the SI based on Eqs. (9.6), (9.8) and (9.9) leads to the greater improvement, followed up closely by the approach using Eqs. (9.7), (9.8) and (9.9) with first task arriving to the pRouter hosting hardware of Type 2.

9.5 Conclusion

The CloudLightning project attempts to address resource management issues associated with hyper-scale cloud deployments. The complexity of these deployments makes the selection of the most suitable resource to host the next service request a very challenging task. Decentralising the decision process is, itself, not sufficient to adequately address this problem, since any distributed collection of resource managers still have to share relevant information and have to together decide on that resource that meets both the service requirements and the business objectives of the cloud service provider. These business objectives can be many and varied, and in the CloudLightning project, they are captured through the use of weighted assessment functions. These functions are used to measure various aspects of the status of the system and this information is subsequently used to determine how close the system is to achieving the business object associated with each respective function. By dynamically weighting assessment functions, the cloud service provider can steer the evolution of the system in the direction of those objects that reflect the providers' immediate needs. As status information is propagated upwards through the CloudLightning hierarchy, it is combined into a view of the underlying levels. This view reflects how well those parts of the system are performing with respect to the business objectives and, by extension, how suitable those levels are to respond to imminent resource requests. The suitability is reflected in a measure known as the suitability index.

An important business object for cloud providers is to minimise energy consumption. The CloudLightning architecture embodies many heterogeneous resources, each with its own energy consumption and exploitation characteristics. The work described here illustrates how the suitability index can be specifically tailored in such a complex environment in support of globally minimising energy consumption. This specially tailored form of the suitability index was evaluated empirically, and the results were presented, showing the improvement of about 6% over the unspecialised suitability index calculation.

The CloudLightning architecture is designed so that it can be easily extended with a multiplicity of heterogeneous resource types. This is achieved using a plug and play mechanism. When a new resource is added to the system using this mechanism, it is assigned to an appropriate resource manager, so that it can be effectively managed. Decision on the most appropriate resource manager to manage a resource is an analogous process to deciding on the most appropriate resource to host a service. Thus, the descriptor, representing a new addition to the resource fabric, follows the path of the lowest (as opposed to the highest) suitability indices until it becomes associated with an appropriate resource manager. This ensures that the resource is placed into the system so as to maximise its utility in meeting the business objectives and in balancing the values of the suitability indices across the cloud.

Acknowledgements This work is funded by the European Union's Horizon 2020 Research and Innovation Programme through the CloudLightning project under Grant Agreement Number 643946.

References

1. Ahuja M, Chen CC, Gottapu R, Hallmann J, Hasan W, Johnson R, Kozyrczak M, Pabbati R, Pandit N, Pokuri S et al (2009) Peta-scale data warehousing at Yahoo! In: Proceedings of the 2009 ACM SIGMOD international conference on management of data. ACM, pp 855–862
2. Barroso LA, Clidaras J, Hölzle U (2013) The datacenter as a computer: an introduction to the design of warehouse-scale machines. In: Synthesis lectures on computer architecture, vol 8, no 3, pp 1–154
3. Beloglazov A, Buyya R, Lee YC, Zomaya A et al (2011) A taxonomy and survey of energy-efficient data centers and cloud computing systems. Adv Comput 82(2):47–111
4. Beloglazov A, Abawajy J, Buyya R (2012) Energy-aware resource allocation heuristics for efficient management of data centers for cloud computing. Future Gener Comput Syst 28(5):755–768
5. Calheiros RN, Ranjan R, Beloglazov A, De Rose CAF, Buyya R (2011) Cloudsim: a toolkit for modeling and simulation of cloud computing environments and evaluation of resource provisioning algorithms. Softw Pract Exper 41(1):23–50. https://doi.org/10.1002/spe.995
6. Dong D, Herbert J (2013) Energy efficient VM placement supported by data analytic service. In: 2013 13th IEEE/ACM international symposium on cluster, cloud and grid computing (CCGrid). IEEE, pp 648–655
7. Etro F (2009) The economic impact of cloud computing on business creation, employment and output in europe. Rev Bus Econ 54(2):179–208
8. Filelis-Papadopoulos C, Xiong H, Spataru A, Castane G, Dong D, Gravvanis G, Morrison JP (2017) A generic framework supporting self-organisation and self-management in hierarchical systems. In: The 16th international symposium on parallel and distributed computing (ISPDC 2017), Paper accepted
9. Hauswald J, Laurenzano MA, Zhang Y, Li C, Rovinski A, Khurana A, Dreslinski RG, Mudge T, Petrucci V, Tang L et al (2015) Sirius: an open end-to-end voice and vision personal assistant and its implications for future warehouse scale computers. In: Proceedings of the twentieth international conference on architectural support for programming languages and operating systems. ACM, pp 223–238
10. Hindman B, Konwinski A, Zaharia M, Ghodsi A, Joseph AD, Katz R, Shenker S, Stoica I (2011) Mesos: a platform for fine-grained resource sharing in the data center. In: Proceedings of the 8th USENIX conference on networked systems design and implementation (NSDI 2011), pp 295–308
11. Joseph E, Conway S, Dekate C, Cohen L (2014) IDC HPC update at ISC14
12. Marinescu DC (2016) Complex systems and clouds: a self-organization and self-management perspective. Morgan Kaufmann
13. Núñez A, Vázquez-Poletti JL, Caminero AC, Castañé GG, Carretero J, Llorente IM (2012) iCanCloud: a flexible and scalable cloud infrastructure simulator. J Grid Comput 10(1):185–209
14. Schubert L, Jeffery K, Neidecker-Lutz B (2010) The future of cloud computing: opportunities for European cloud computing beyond 2010. Expert Group report, public version 1
15. Sohrabi S, Moser I (2015) A survey on energy-aware cloud. Eur J Adv Eng Technol 2(2):80–91
16. Sverdlik Y (2014) Survey: industry average data center PUE stays nearly flat over four years. Data Center Knowl 2(06)
17. Tang L, Mars J, Zhang X, Hagmann R, Hundt R, Tune E (2013) Optimizing Google's warehouse scale computers: the NUMA experience. In: 2013 IEEE 19th international symposium on high performance computer architecture (HPCA2013). IEEE, pp 188–197
18. Tian W, Xu M, Chen A, Li G, Wang X, Chen Y (2015) Open-source simulators for cloud computing: comparative study and challenging issues. Simul Model Pract Theory (special issue on Cloud Simulation) 58(Part 2):239–254
19. Whitney J, Delforge P (2014) Data center efficiency assessment. National Resources Defense Council, New York

Chapter 10
Developing Low-Power Image Processing Applications with the TULIPP Reference Platform Instance

Tobias Kalb, Lester Kalms, Diana Göhringer, Carlota Pons, Ananya Muddukrishna, Magnus Jahre, Boitumelo Ruf, Tobias Schuchert, Igor Tchouchenkov, Carl Ehrenstråhle, Magnus Peterson, Flemming Christensen, Antonio Paolillo, Ben Rodriguez and Philippe Millet

10.1 Introduction

In today's industry, an increasing amount of applications relies on vision-based techniques. The resulting image processing systems cover a wide field of application, examples being medical imaging, automotive advanced driver assistance systems (ADAS) and unmanned aerial vehicles (UAVs). All these applications demand high computing performance, yet strict requirements and constraints of an embedded system have to be met. Therefore, both the embedded system and the image processing

T. Kalb (✉)
Ruhr-University Bochum, Bochum, Germany
e-mail: tobias.kalb@rub.de

L. Kalms · D. Göhringer
TU Dresden, Dresden, Germany
e-mail: lester.kalms@tu-dresden.de

D. Göhringer
e-mail: diana.goehringer@tu-dresden.de

C. Pons
Efficient Innovation, Castelnau-le-Lez, France
e-mail: c.pons@efficient-innovation.com

A. Muddukrishna · M. Jahre
Norwegian University of Science and Technology, Trondheim, Norway
e-mail: ananya.muddukrishna@idi.ntnu.no

M. Jahre
e-mail: magnus.jahre@idi.ntnu.no

B. Ruf · T. Schuchert · I. Tchouchenkov
Fraunhofer Institute of Optronics, System Technologies and Image Exploitation IOSB, Karlsruhe, Germany
e-mail: boitumelo.ruf@iosb.fraunhofer.de

T. Schuchert
e-mail: tobias.schuchert@iosb.fraunhofer.de

C. Kachris et al. (eds.), *Hardware Accelerators in Data Centers*,
https://doi.org/10.1007/978-3-319-92792-3_10

application have to be optimized. Embedded image processing systems are expected to be a ubiquitous part of our society with a direct connection to cloud computing, servers and data centres. Today, the complexity of systems is continuously growing. The design and implementation of an image processing platform will be an even more challenging task for developers. A non-optimized design cannot satisfy the demand for low power and high performance.

Embedded platforms using a single processor do not offer enough processing power to run modern image processing applications. On the contrary, the number of sensors is growing and the system simultaneously has to deal with increasing connectivity and data, which puts even higher requirements on the processing of data and communication. The demands of modern and future low-power computing systems are not met by current architectures. Embedded vision systems are often battery powered. Therefore, an efficient architecture is needed so that performance and energy efficiency results in longer battery life and better user experience. Resource-intensive image processing can be offloaded to cloud computing or data centres, but this compromises real-time constraints and latencies of the application. Thus, the embedded system has to deliver low-power yet high-performance computing for image processing.

Regarding embedded platforms, there are two approaches to reach the performance requirements of current image processing algorithms. The first approach is to run the image processing applications on low-power graphics processing units (GPUs). This can significantly improve processing power. The second approach offers the possibility to further reduce energy consumption by using field programmable gate arrays (FPGAs) for image processing. Using state-of-the-art technology, both GPU and FPGA are connected to an embedded processor forming a system-on-chip (SoC). Heterogeneous architectures offer promising features for

I. Tchouchenkov
e-mail: igor.tchouchenkov@iosb.fraunhofer.de

C. Ehrenstråhle · M. Peterson
Synective Labs AB, Göteborg, Sweden
e-mail: carl.ehrenstrahle@synective.se

M. Peterson
e-mail: magnus.peterson@synective.se

F. Christensen
Sundance Multiprocessor Technology Ltd., Chesham, UK
e-mail: flemming.c@sundance.com

A. Paolillo · B. Rodriguez
HIPPEROS S.A., Ottignies-Louvain-la-Neuve, Belgium
e-mail: antonio.paolillo@hipperos.com

B. Rodriguez
e-mail: ben.rodriguez@hipperos.com

P. Millet
Thales, Toulouse, France
philippe.millet@thalesgroup.com

modern vision-based applications. Each type of processing element has a type of processing for which it performs best. To get the best out of an efficient heterogeneous hardware platform, it is important to efficiently map the application onto the different processing elements and manage its execution at runtime. The increasing demand for vision-based systems also asks for reducing time-to-market, development and rework costs on a product as well as maximizing reuse of designs. However, there is still no standard for high-performance embedded computing systems on heterogeneous platforms in the domain of vision-based systems.

TULIPP [1, 2] aims to push forward a reference platform defining implementation rules to provide designers with guaranteed high-performance solutions for vision-based systems. TULIPP also considers flexibility and scalability for applications demanding more processing power. Therefore, the platform will be set up for heterogeneous multicore and multiprocessor architectures. Several instances of the platform can be combined to further increase processing performance. During the project, a heterogeneous hardware reference platform, a real-time operating system (RTOS) and a productivity-enhancing set of development utilities will be developed. These three components will enable high-performance image processing for modern low-power embedded systems and, in addition, will allow validating the implementation rules and guidelines. The TULIPP project will publish a reference platform handbook, which allows developers to easily follow and apply the implementation rules to find an optimal solution for their image processing application. Thus, TULIPP-compliant custom platforms and applications can be designed at reduced development time and costs.

10.2 The TULIPP Use Cases

The development of the TULIPP reference platform as well as the TULIPP reference platform handbook are use case driven. The three use-case scenarios feature applications for medical imaging, for automotive systems and for UAVs. The applications cover different scenarios, yet the requirements for the embedded systems as such are similar. All three use cases have to operate their image processing applications with high computational performance, low power consumption as well as reduced overall system size and weight. In addition, these applications have to comply with short deterministic latencies, which requires to process images within real-time constraints. Using these three use-case scenarios, the developments of the project, namely the hardware platform, the real-time operating system, and the productivity-enhancing utilities, can be evaluated at the best to provide optimal implementation rules and guidelines for the TULIPP reference platform handbook.

10.2.1 Medical X-Ray Imaging

In medical imaging, mobile equipment is expected to replace high-end infrastructure devices. Modern-day surgery requires that the surgeon has precise control of their movements and at times is able to see the path that blood flows through veins and arteries. Complex imaging systems have to be used to achieve this.

Dedicated to X-ray instruments, the work of the TULIPP project is highly relevant to a significant part of the market share, in particular through its Mobile C-Arm use case, which is a perfect example of a medical system that improves surgical efficiency. In real time, during an operation, this device displays a view of the inside of a patient's body, allowing the surgeon to make small incisions rather than larger cuts and to target the region with greater accuracy. This leads to faster recovery times and lower risks of hospital-acquired infection. Current X-ray sensors are able to provide live images and video in real time. The drawback of this is the radiation dose: 30 times what we receive from our natural surroundings each day. This radiation is received not only by the patient but also by the medical staff, week in, week out.

While the X-ray sensor is very sensitive, lowering the emission dose increases the level of noise on the pictures, making it unreadable. This can be corrected with proper processing.

From a regulatory point of view, the radiation that the patient is exposed to must have a specific purpose. Thus, each photon that passes through the patient and is received by the sensor must be delivered to the practitioner; no frame should ever be lost. This brings about the need to manage side by side strong real-time constraints and high-performance computing.

The medical use case of the TULIPP project deals with this problem in real-time X-ray image processing. The goal is to reduce the radiation dose to much safer levels while keeping the image quality and the required latency and rate. The image processing is done by an embedded system within a mobile device called 'C-arm' because of its 'C' shape. The X-ray source is located at one end of the C and the sensor at the other end. This shape allows to have the patient in the middle and to take images of any of its body parts through any angle and direction. The C-arm is used during surgery and delivers an X-ray video stream in real time. The chosen algorithms reduce noise from the sensors and enhance the image in order to provide enough details to the surgeon. The algorithms require high-performance processing.

In the TULIPP project, our aim is to try to reduce the level of radiation by 75%. As a result of this, more powerful image processing will be required in order to still be able to see small details in the human body that are crucial during surgery. Since most operating theatres are confined environments crowded with staff and equipment, the device needs to be small and mobile. A system that integrates the processing close to the sensor is ideal to help reduce extraneous wires and improves the mobility of the equipment. The system needs to be compact but also has a low power draw since heat and other RF emissions could disturb the sensors and eventually actually add more noise to the signal. When we add up the hard real-time constraints to which the system must comply due to part of regulatory constraints regarding devices used in medical environments, this combination of requirements makes this use case a challenge to design and develop a matching solution (Fig. 10.1).

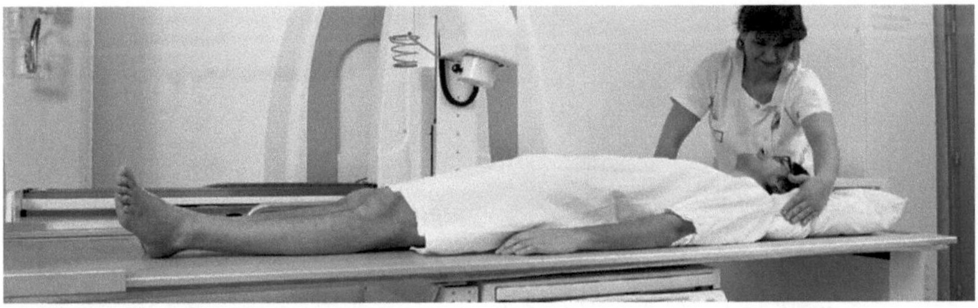

Fig. 10.1 Digital radiography

10.2.2 Advanced Driver Assistance

In the automotive domain, more electronic devices are going to be integrated into cars in the future. One of the most promising segments for embedded vision systems are the advanced driver assistance systems (ADAS). A steep growth is expected for the next five to ten years. Here, the automotive industry puts a strong focus on driving safety and pedestrian safety with vision-based systems as one of the enablers for many new and innovative solutions. This includes both passive and active safety systems. The most interesting fields of application for embedded vision systems include vehicle, pedestrian and object detection, traffic sign recognition, lane detection, night vision, surround view and driver monitoring. Data from optical sensors is often combined with data from other sensors to either guide or assist the driver, or to take control of the vehicle by automatic braking, automatic lane keeping, park assist, etc. These applications will over the years be refined and enhanced, resulting in fully autonomous driving solutions some 10 years from now.

The automotive use case of the TULIPP project is focused on pedestrian detection. The purpose of pedestrian detection algorithms is to recognize humans in an image collected by an optical sensor. Detected pedestrians can then be used to trigger further processing, e.g. automated braking. In TULIPP, the implemented algorithm performs the pedestrian detection by feeding a set of trained classifiers various processed forms of an input image.

ADAS vision systems require real-time, low-latency processing, at high to very high computational load. They need to be robust and reliable, and will often be treated as safety critical systems. The TULIPP project addresses all these questions. By offering a toolset and standardization, it will help the designers to focus on the image processing application rather than platform details. The TULIPP ADAS use case shows how a typical automotive vision application, pedestrian detection, can be facilitated by the TULIPP platform and how characteristics like low power, high performance and robustness are natively supported (Fig. 10.2).

Fig. 10.2 ADAS object recognition

10.2.3 Autonomous UAVs

In the domain of unmanned aerial vehicles (UAVs), onboard image processing in real time is a key technology for autonomous operation [3]. Small UAVs have entered a large range of applications as their underlying technology has improved. This also allowed for the exploration of a new and large range of applications. Today, applications for surveillance, search and rescue, video production, logistics and research are just a small subset of possible scenarios [4]. With the growing amount of UAVs, however, the number of crashes and problems with controlled operation is increasing. Problems can be caused by several sources, e.g. operator error as well as mechanical or electrical malfunction. In a worst-case scenario, this error not only involves the UAV itself but also humans, goods or infrastructure [5]. Therefore, UAVs need more intelligent control and interaction systems, such as automatic collision avoidance or more robust pose estimation, to minimize risks of failure.

The goal of the UAV use case in TULIPP is to estimate depth images from a stereo camera set-up. Orientated in the direction of flight, the depth images are used to detect objects in the path of the UAV. In further stages, detected object is then used for collision avoidance. This approach creates a more autonomous and more intelligent solution. The problem is that more intelligence needs more computing power, which is very limited especially on small UAVs. Yet, image processing has to be onboard in real time, also considering weight and power constraints.

The TULIPP solution aims to fill this processing gap by using its good performance-to-weight and power consumption-to-weight figures. We aim to use computer vision algorithms such as stereo and depth estimation to detect obstacles and evaluate the

Fig. 10.3 Autonomous unmanned aerial vehicle

surroundings in order to make the UAV more intelligent. For this purpose, we attach the TULIPP reference platform with a stereo camera set-up orientated in direction of flight to a UAV. Our goal is to use stereo algorithms to automatically detect obstacles in real time that are within dangerous vicinity in front of the UAV and to avoid a collision. Therefore, we will provide a processing chain which is quite common to any stereo vision-based application. The processing chain contains stereo image acquisition, preprocessing, like image rectification, a depth estimation algorithm based on semi-global matching, similar to [6], an obstacle avoidance algorithm as well as an interface to an external system (UAV) (Fig. 10.3).

These three use cases represent an ideal combination of applications for the TULIPP project. Each use case requires embedded high-performance low-power image processing, but the constraints differ for each use-case scenario. Striving for optimization for each use case, the developments of the TULIPP project will then provide a flexible and extensible solution including rules and definitions for the hardware platform, the operating system and the utilities used to design, develop and deploy an embedded high-performance low-power image processing application.

10.3 The TULIPP Reference Platform Instance

The TULIPP project is developing and will provide a reference platform. The reference platform is presented in the context of the starter kit, a conceptual package consisting of a platform instance, project applications and the reference platform handbook. The aim of the starter kit is to provide engineers with a generic evaluation platform that serves as a base for productively developing low-power image processing applications. The platform instance is a physical processing system consisting of hardware, Operating System (OS), and application development tools. This platform instance demonstrates the results of the project using the applications of the presented use cases.

The reference platform handbook is a set of guidelines for low-power image processing embedded systems. We use guidelines as shorthand for the reference platform handbook. Guidelines recommend application implementation methods supported by the platform instance. A guideline is a goal-oriented, expert-formulated encapsulation of advice and recommended implementation methods for low-power image processing. A vendor platform that enables guidelines by providing suitable implementation methods is called an instance. An instance is fully compliant if it provides recommended implementation methods for all the guidelines that it supports. We envisage that compliance with guidelines will be judged and certified by an independent body identified by the ecosystem of stakeholders.

The reference platform is used to define implementation rules and interfaces to tackle power consumption for high and efficient computing performance demands for image processing applications. The main objective is to provide a new approach to find an optimal solution for a vision-based system. The complex task of designing and evaluating different, interleaving and evolving hardware and software components is then eased so that the overall cost of image processing devices will be reduced drastically. The universal and well-defined interfaces of the reference platform offer the possibility to include new generations of hardware devices and software components without significant overheads and costs for redesigning the system. It allows developers to efficiently design more embedded and less power consuming image processing platforms.

The TULIPP image processing solution aims for scalable high performance and mechanical flexibility to be able to comply with heat dissipation and size constraints, low cost and low power consumption. The inherent idea of customization guides the activities of TULIPP to set up guidelines and definitions on how to use and combine heterogeneous technology at its best. Thus, an optimal solution—in terms of performance, energy efficiency and development costs—for a customized image processing system can be found. Furthermore, TULIPP takes into consideration that the reference platform will evolve at the same pace as modern technology. This ensures that developers can benefit from the improvements future technologies and devices offer but it is a challenge at the same time because the reference platform has to assimilate new technology as it comes up.

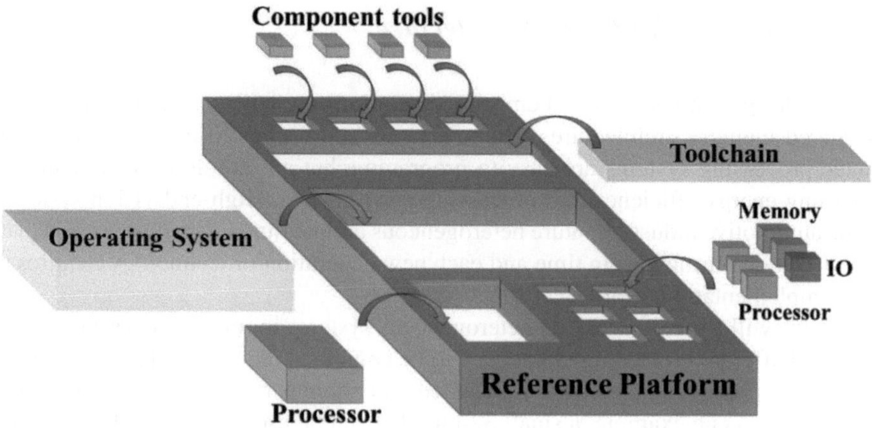

Fig. 10.4 TULIPP reference platform

Instead of developing a generic platform solution that should fit all the applications, TULIPP proposes to focus its efforts on developing a guide for a reference platform that helps the designer in making choices for the components and interfaces. This reference platform is a versatile ideal platform described through a set of guidelines. By following the guidelines, one can implement features to tackle power consumption while delivering high and efficient computing performance for image processing applications under real-time constraints on processing rate and latencies. This solution is much more beneficial and future-proof including all aspects of the development of an image processing platform. TULIPP will build the reference platform through industrial consensus dedicated to low-power real-time image processing applications (Fig. 10.4).

The project will concentrate on interfaces between the components of the platform (hardware, utilities, operating system and middleware libraries) as well as design and implementation processes. Following the guidelines, a developer will be able to produce a compliant platform and benefit from the technological advances generated by the project. In addition, vendors will be able to produce a compliant part and plug it into an existing platform. Thus, a TULIPP-compliant platform instance benefits both to the developers and the vendors. To achieve this goal, the TULIPP project will set up and work closely with an ecosystem formed with platform-part providers (e.g. chip, processing board, operating system, processing libraries, toolchain, etc.) and application developers. This allows incorporating valuable feedback during the project lifetime.

10.3.1 TULIPP Hardware Architecture

Current designs in vision-based embedded solutions are built on single-core CPUs or shared memory architectures. Homogeneous approaches to modern embedded image processing systems are easy to programme but are not an optimal solution regarding energy efficiency and processing performance. High-end vision systems in the automotive industry feature heterogeneous architectures. The drawback is that each iteration during design time and each new generation of technology requires a huge implementation effort.

TULIPP will focus its work on heterogeneous systems in image processing applications. Different processing elements will be combined in a hardware architecture, where each processing element is best suited for certain parts of an image processing application. As an example, a small 32-bit CPU can be used for controlling in- and outputs. The processing of images can then be executed on multiple 64-bit CPUs with additional acceleration by FPGAs or embedded GPUs.

The TULIPP reference platform provides a template for heterogeneous computing architectures and systems. Modern SoCs demonstrate the potential of heterogeneous systems. The NVIDIA Tegra-K1 [7] provides high performance by combining an ARM processor with a GPGPU. Similarly, Xilinx puts great efforts on the Zynq devices. Here, an ARM processor is combined with an FPGA. Xilinx UltraScale+ MPSoCs takes one step further and combines 64-bit and 32-bit ARM architectures together with dedicated real-time cores and an FPGA [8]. The goal of the TULIPP reference platform is not only to optimally utilize a single SoC. Moreover, the project aims to connect different SoCs. Different parts of an image processing application can then be run on the best-suited computing architecture. An adaptive system allows running an application energy efficient yet high performant. The hardware platform will be fine-tuned and configured for each application. Therefore, TULIPP defines how to select SoCs suitable to build a TULIPP platform instance and how to efficiently interconnect several SoCs. Switch-off mechanisms, adjustable operating frequencies and dynamic partial reconfiguration (DPR) [9] further reduce the cost of unused system resources during runtime.

As of the time of writing, TULIPP uses Xilinx Zynq SoCs to thoroughly test different combinations of hardware and interfaces on- and off-chip. The inherent heterogeneity of these devices is used to derive definitions, implementation rules and guidelines for the TULIPP reference platform and the TULIPP reference platform handbook. The TULIPP hardware architecture uses the PC/104 form factor. This form factor is already supported by many vendors, which are committed to the ongoing development of this specification [10]. It is mature, modern, open standard and expandable but also capable for stand-alone applications. These features are perfectly suited for a scalable and heterogeneous TULIPP hardware platform.

10.3.2 TULIPP Operating System and Low-Level Libraries

Today, there is a significant gap between research and commercial implementation of an RTOS. Research results feature scheduling algorithms, resource sharing algorithms and inter-process communication (IPC) protocols. Yet these innovations are rarely incorporated in commercial systems. Existing RTOS are mainly targeted towards single-core designs, and multicore approaches focus on homogeneous SMP architectures. In addition, the need for energy efficiency is rarely supported. This includes, for example, power-aware scheduling at the kernel level [11]. As a result, system lifetime and reliability are reduced [12].

In TULIPP, the design of the operating system and low-level libraries is targeted towards low power consumption and image processing. The RTOS kernel supports heterogeneous architectures and power-aware features. Communication and synchronization mechanisms are implemented so that the operating system correctly operates on the instantiated processing elements of the hardware platform. The footprint—i.e. the binary size—of the RTOS is kept small for hardware components only embedding a small local memory. Frequent accesses to bigger memories like DDR are not suitable for running image processing applications in real time and consume more energy. The TULIPP reference platform provides primitives so that components can be integrated or wrapped in the low-level library available for the programmer. In addition, standard APIs will be modified and extended to comply with the requirements of low power consumption and high-performance image processing. Thus, extensions and modified standards are proposed as a pre-norm.

The RTOS in TULIPP is developed to efficiently handle heterogeneous hardware processing resources of multicore CPUs and FPGAs with a strong focus on finding the right balance between low power footprint and high computing performance.

The RTOS solution proposed by the TULIPP project is a new master–slave microkernel architecture specifically designed for heterogeneous multicores. It features a small footprint, low power consumption and good scalability. This is a combination of several features. Power-aware schedulers—i.e. extended earliest deadline first (EDF) instead of rate monotonic (RM) schedulers [13]—reduce the overall power consumption of applications. Moreover, it has been shown in the real-time literature that schedulers based on a parallel task model are well suited to be extended to power-aware scheduler [14]. Therefore, the TULIPP platform includes an RTOS capable of scheduling parallel real-time tasks (software or hardware) associated with the right runtime libraries allowing to easily design parallel workload to be run on the different heterogeneous components of the target platform. This, combined with optimizations provided by the offline utilities of TULIPP presented in the next section will result in an image processing embedded system suited to the user requirements in terms of power consumption and computing performance.

The hard real-time scheduling of hardware/software tasks is combined with virtual memory management to isolate processes and efficient IPC mechanisms to allow these processes to communicate. Developers can then use the reliable real-time guarantees and easy programmability of the provided RTOS for optimized low-power image processing applications.

The TULIPP real-time operating system is designed by HIPPEROS and based on its family of RTOSes [15, 16]. It supports standard tools to interface with hardware (bootloaders, debuggers, etc.), and low-level libraries shipped with the operating system support APIs validated for low-power embedded image processing applications (POSIX, OpenCV, OpenMP, etc.) [17]. The operating system interfaces are developed to be optimally integrated with the TULIPP hardware platform and the supporting development toolchain and utilities. Additionally, the RTOS environment is adapted to suit the modern advantages of the heterogeneous platforms [18]. For example, for heterogeneous platforms as the Zynq device, the HIPPEROS RTOS provides APIs to partially reconfigure at runtime what is running in the FPGA, enabling the Xilinx dynamic partial reconfiguration feature and exposing it to the image processing application developer (the user of the TULIPP platform).

Thus, the main advantage of the TULIPP RTOS solution is its efficiency for heterogeneous parallelism and power optimization.

10.3.3 TULIPP Toolchain

The current state of the art in development tools for heterogeneous image processing systems requires a lot of interaction and experience with several different vendor-specific tools. Each component is accompanied by its own complex tools. The developer has to spend a significant amount of time to master these tools. This results in low productivity and a reduced innovation rate. Extensive reviews of such heterogeneous system development tools exist [9, 19, 20]. A TULIPP-compliant platform can feature hardware components from several different vendors. In general, each component is supplied with its own specific toolchain and Integrated Development Environment (IDE). To efficiently develop low-power, high- performance applications on TULIPP hardware platforms, the programmer needs to gain expertise in several tools. This inhibits productivity, as such expertise takes a long time to develop. In addition, lack of knowledge or expertise in a particular vendor—device or tool—may also prevent developers from selecting hardware components, which would be best suited for the power and performance requirements of a specific image processing application.

TULIPP proposes a solution using toolchain utilities that allow developers to use multivendor tools more efficiently and productively. The focus of the utilities is put on an improved system set-up, its analysis and optimization. This includes mapping an application optimally onto a heterogeneous, TULIPP-compliant platform. The set of utilities is called STHEM—supporting utilities for heterogeneous embedded image processing. STHEM wraps around, extends and connects existing vendor tools to present a seamless mapping and performance/energy analysis interface to programmers. Developers are able to map parts of an application onto suitable components using STHEM interfaces. Primitives and library routines can be used to handle control and communication of the components. Problem areas of the application can be

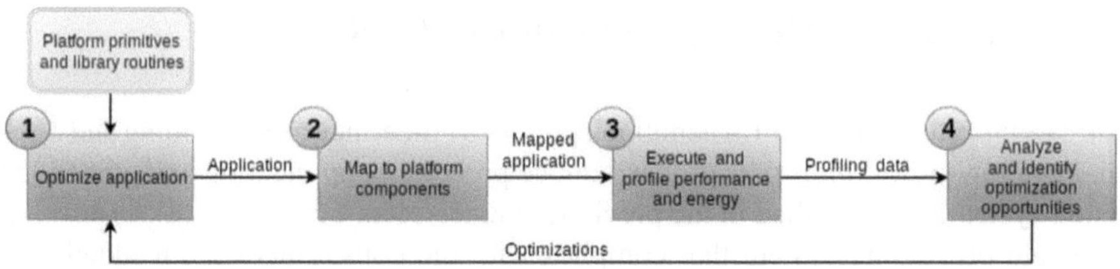

Fig. 10.5 High-level overview of the iterative workflow using STHEM

identified with STHEM's utilities for performance and energy consumption analysis. Furthermore, the utilities also identify optimization opportunities. Thus, the developer is guided towards an energy-efficient and high-performant image processing application and platform.

An overview of the workflow is given in Fig. 10.5. STHEM provides efficient usability of expert-written mechanisms for the improvement of an application. The workflow is iterated until desired performance and energy profile is reached. In the first stage of the workflow, programmers write application code. The optimizations identified in the previous iteration are also applied in this first stage. Programmers are also assisted with platform-specific primitives and library routines that abstract away commonly used domain-specific functionality. In the second stage, developers are supported in the mapping of an application to components using easy-to-use, expert-written mapping directives. The third stage is used for platform configuration, application execution and profiling. Analysis of the execution and profiling is provided in the fourth stage. Visualizations are used to highlight profiling data and problem areas [21]. In addition, STHEM suggests optimization strategies to programmers and offers to automatically explore the design space.

STHEM is built as an Eclipse 4 RCP plugin to facilitate an integrated workflow with popular vendor tools that integrate into the Eclipse IDE [22–24]. The first implementation of STHEM is designed for Xilinx SDSoC [22]. Thus, all the components of the TULIPP reference platform—heterogeneous hardware platform, multicore RTOS and supporting utilities—can be evaluated and tailored towards optimal energy efficiency and high computing performance for image processing applications.

With the three components described above—hardware architecture, operating system and toolchain—the TULIPP project aims to provide rules and definitions for a reference hardware platform. Thus, the TULIPP project does not only provide an extensible set of components for embedded high-performance low-power image processing applications. Moreover, the insights and knowledge gained by developing these components is comprehensively written down in the TULIPP reference platform handbook.

10.4 The TULIPP Reference Platform Handbook

The TULIPP reference platform handbook provides guidelines on developing and optimizing a low-power high-performance image processing application and embedded vision system. As described in the previous chapters, it is a complex task to develop an image processing system that complies with the constraints of embedded systems and the demand for low power consumption and high computing performance. Embedded vision systems feature a complex combination of different design steps, including component selection and platform set-up, application and algorithm optimization as well as application mapping.

The guidelines provided by the TULIPP reference handbook support the developer by managing the complexity of designing modern low-power image processing embedded systems. Therefore, guidelines are derived from the expert knowledge gained during the development of the TULIPP reference platform. This knowledge is split into advices and recommended implementation methods. Advices support the developer on what to do so that an image processing application and platform can be optimized. The recommended implementation methods describe in detail how to achieve the suggested optimizations.

The guidelines of the handbook consider all the components of the TULIPP reference platform. Developers are supported in the selection of TULIPP-compliant hardware components and operating system settings best suited for their application. In addition, guidelines are formulated on how to rewrite the application and how to use APIs, libraries and pragmas achieving a more efficient image processing application. The guidelines always put the focus on low power consumption, high and heterogeneous computing performance and real-time image processing.

As a long-term goal, the TULIPP handbook aims to improve productivity as well as influence new standards for heterogeneous embedded vision systems operating under real-time constraints with low power consumption and high computing performance. The expert insights of the guidelines then not only support developers by optimizing their application, but also vendors by providing TULIPP-compliant hardware and software. For developers, this has the potential to significantly improve productivity. Costly mistakes are avoided and vast design and implementation spaces are pruned. Vendors are encouraged to strive for compliance by providing suitable hardware and software. Thus, the efforts of the TULIPP project pave the way for future standards in embedded low-power image processing systems.

10.5 Future Goals and Outlook

The main objective of the TULIPP project is to develop a reference platform for high-performance and energy-efficient embedded systems for image processing applications. In addition to this reference platform, TULIPP is going to support developers with a reference platform handbook. The handbook provides guidelines on developing and optimizing low-power high-performance embedded vision systems.

To achieve its goals, the TULIPP project develops and provides a starter kit. This Starter Kit is used to demonstrate the potential of the projects' approaches to the design, development and implementation of embedded vision systems. The TULIPP starter kit allows the project to track its efforts and demonstrate its use cases. Furthermore, the kit also allows developers to design high-performant yet low-power image processing platforms and applications. The starter kit will feature a first version of the TULIPP reference platform and handbook. The hardware component consists of a PC/104 board with a small system-on-module (SoM). Here, the SoM features a Xilinx Zynq device for heterogeneous computing performance. The toolchain component consists of a collection of application analysis utilities referred to as support utilities for heterogeneous embedded image processing (STHEM). STHEM will augment existing vendor toolchains by automating analysis procedures and supporting more efficient application design. This reduces time-to-market, improves developer productivity and system quality by making it easier and faster to arrive at an implementation that meets the requirements. The hardware platform and STHEM will be completed with the HIPPEROS real-time operating system and its low-level libraries, allowing to design and run highly efficient embedded vision-based applications and platforms.

The reference platform and handbook cover all the aspects of designing embedded systems for vision-based applications: from computing hardware, operating system and low-level libraries to programming toolchain and utilities. The reference platform also defines a set of interfaces between basic components and implementation rules to facilitate prospective system design. The implementation rules and guidelines of the reference platform handbook support developers designing embedded image processing platforms. This significantly contributes to a reduction of time-to-market and development costs. The TULIPP project is going to leverage the utilization of heterogeneous embedded computing platforms for image processing applications. Thus, the TULIPP project aims to pave the way for standards for embedded high-performance low-power image processing in industrial applications.

The TULIPP project establishes a valuable advisory board and ecosystem. The advisory board consists of leading experts from industry and academia with a strong focus on image processing applications and platforms targeted towards embedded systems. Thus, valuable feedback is used during the project to further leverage and promote its developments. Incorporating high-quality feedback and utilizing its developments and guidelines the TULIPP project puts great efforts into industrial usage and standardization of ubiquitous low-power embedded systems for image processing platforms. The state and progress of the project will be available to the public at the website of the TULIPP project [1].

Acknowledgements The project is funded by European Commission under the H2020 Framework Programme for Research and Innovation under grant agreement No 688403.

References

1. Tulipp (2017) TULIPP: towards ubiquitous low-power image processing platforms—high, efficient and guaranteed computing performance for image processing applications. http://www.tulipp.eu. Accessed on 11 May 2017
2. Kalb T et al (2016) TULIPP: towards ubiquitous low-power image processing platforms. In: Proceedings of the international conference on embedded computer systems: architectures, modeling and simulation (SAMOS XV)
3. Sung C-K, Segor F (2012) Onboard pattern recognition for autonomous UAV landing. In: Proceedings of the SPIE 8499, applications of digital image processing XXXV, 84991K
4. Kuntze HB et al (2012) SENEKA–sensor network with mobile robots for disaster management. In: 2012 IEEE conference on technologies for homeland security (HST). Waltham, MA, pp 406–410
5. Tchouchenkov I, Segor F, Schoenbein R, Kollmann M, Bierhoff T, Herbold M (2016) Detection and protection against unwanted small UAVs. In: Proceedings of the eleventh international conference on systems ICONS
6. Ruf B, Schuchert T (2016) Towards real-time change detection in videos based on existing 3D models. In: Proceedings of SPIE, Edinburgh, UK, vol 10004, pp 100041H–100041H-14
7. NVIDIA Corporation (2014) Whitepaper: NVIDIA Tegra K1—a new era in mobile computing. https://ww.nvidia.com/content/PDF/tegra_white_papers/tegra-K1-whitepaper.pdf. Accessed on 28 April 2017
8. Xilinx Inc. (2015) White Paper: Zynq UltraScale+ MPSoCs—unleash the unparalleled power and flexibility of Zynq UltraScale+ MPSoCs (WP470). http://www.xilinx.com/support/documentation/white_papers/wp470-ultrascale-plus-power-flexibility.pdf. Accessed on 28 April 2017
9. Cong J, Liu B, Neuendorffer S, Noguera J, Vissers K, Zhang Z (2011) High-level synthesis for FPGAs: from prototyping to deployment. IEEE Trans Comput Aided Design Integr Circuits Syst 30(4):473–491
10. PC/104 Consortium (2017) PC/104 Consortium—supporting legacy technology while developing new solutions for the future PC/104 Consortium. http://pc104.org. Accessed on 11 May 2017
11. Mentor Graphics (2017) Nucleus RTOS. https://www.mentor.com/embeddedsoftware/nucleus/. Accessed on 04 May 2017
12. Brandenburg B (2011) Scheduling and locking in multiprocessor real-time operating systems. PhD dissertation, The University of North Carolina
13. Blazewicz J, Ecker KH, Pesch E, Schmidt G, Weglarz J (2001) Scheduling computer and manufacturing processes. Springer, Berlin. ISBN 3-540-41931-4
14. Paolillo A, Goossens J, Hettiarachchi PM, Fisher N (2014) Power minimization for parallel real-time systems with malleable jobs and homogeneous frequencies. In: The 20th IEEE international conference on embedded and real-time computing systems and applications, Chongqing, China, August 2014
15. Paolillo A, Desenfans O, Svoboda V, Goossens J, Rodriguez B (2015) A new configurable and parallel embedded real-time micro-kernel for multi-core platforms. In: Proceedings of the ECRTS workshop on operating systems platforms for embedded real-time applications (ECRTS-OSPERT'15), July 2015
16. HIPPEROS S.A. http://www.hipperos.com. Accessed on 09 May 2017
17. Paolillo A, Rodriguez P, Veshchikov N, Goossens J, Rodriguez B (2016) Quantifying energy consumption for practical Fork-Join parallelism on an embedded real-time operating system. In: The 24th ACM international conference on real-time networks and systems, Brest, France, October 2016
18. Martin C, Antonio P, Goossens J, Rodriguez B (2017) Research and implementation challenges of RTOS support for heterogeneous computing platforms. In: HARTS-ULB, Brussels, Belgium, May 2017

19. Jeffers J, Reinders J (2015) High performance parallelism pearls volume two: multicore and many-core programming approaches. Morgan Kaufmann
20. Mittal S, Vetter JS (2015) A survey of CPU-GPU heterogeneous computing techniques. ACM Comput Surv (CSUR) 47(4):69
21. Muddukrishna A, Jonsson PA, Podobas A, Brorsson M (2016) Grain graphs: OpenMP performance analysis made easy. In: Proceedings of the 21st ACM SIGPLAN symposium on principles and practice of parallel programming, New York, NY, USA. pp 28:1–28:13
22. Xilinx Inc. (2017) SDSoC environment user guide (UG1027). http://www.xilinx.com/support/documentation/sw_manuals/xilinx2015_4/ug1027-sdsoc-user-guide.pdf. Accessed on 09 May 2017
23. NVIDIA Nsight|NVIDIA. http://www.nvidia.com/object/nsight.html. Accessed on 09 May 2017
24. Profiling OpenMP* applications with Intel® VTune™ Amplifier XE|Intel® Developer Zone. https://software.intel.com/en-us/articles/profiling-openmp-applications-with-intel-vtune-amplifier-xe. Accessed on 27 May 2015

Chapter 11
Energy-Efficient Heterogeneous Computing at exaSCALE—ECOSCALE

Konstantinos Georgopoulos, Iakovos Mavroidis, Luciano Lavagno, Ioannis Papaefstathiou and Konstantin Bakanov

11.1 Introduction

In order to sustain the ever-increasing demand for storing, transferring and processing data, HPC servers need to significantly improve their efficiency. Scaling the number of cores alone is no longer a feasible solution due to the increasing utility costs and power consumption limitations. Furthermore, while current HPC systems can offer petaflop performance, their architecture limits their capabilities in terms of scalability and energy consumption. Extrapolating from top modern HPC systems, such as China's Tianhe-2 Supercomputer, we estimate that sustaining exaflop performance requires a highly significant 1 GW of power. Similar, albeit smaller, figures are obtained by extrapolating even the best system of the Green 500 list as an initial reference.

Apart from improving transistor and integration technology, important refinements in HPC application development and HPC architecture design are also needed. The ECOSCALE project [21] meets these challenges with a novel and holistic approach for exascale computing, combining a hybrid many-core+OpenCL programming environment with a hierarchical architecture, an intelligent runtime system and middleware and hardware support for sharing distributed and reconfigurable accelerators.

This paper is separated into a number of different sections and subsections so that all important aspects of the ECOSCALE solution can be addressed sufficiently, such as system architecture, runtime and programming model.

K. Georgopoulos · I. Mavroidis (✉) · L. Lavagno · I. Papaefstathiou · K. Bakanov
Telecommunication Systems Institute (TSI), Technical University of Crete, Campus Kounoupidiana, Chania, Crete, Greece
e-mail: jacob@ics.forth.gr

K. Georgopoulos
e-mail: kgeorgopoulos@isc.tuc.gr

C. Kachris et al. (eds.), *Hardware Accelerators in Data Centers*,
https://doi.org/10.1007/978-3-319-92792-3_11

11.2 HPC Application Characteristics

The ECOSCALE architecture has been envisaged such that it suits the characteristics and trends of future HPC applications so that they efficiently scale to exaflop performance. In our attempt to predict future HPC applications, we envision that they will have the following fundamental characteristics:

1. Massive parallelism: Applications can be partitioned into many parallel tasks or threads that can run in parallel. A 1000× increase in today's concurrency will be necessary to achieve exascale throughput [7].
2. Data locality: HPC applications should be characterised by spatial and temporal locality in order to scale. The data accessed by an HPC application will be partitioned in memory sub-domains in such way that memory transfers between these sub-domains are infrequent and efficient.

Subsequently, partitioning into sub-domains will dictate the mapping of tasks to cores and accelerators, so that each task executes on a device that is directly *attached* to the data that the task accesses. Instead of a flat partitioning of the application domain, we foresee that future large-scale HPC applications will perform hierarchical and topological partitioning [16], such as a high-radix Dragonfly or Slim fly topology, of their data into domains. This shall reduce communication distance and latency. A leaf node in this partitioning would correspond to a data domain that fits in the local memory of a single processor–accelerator bundle, which we have coined as *Worker*. Moving up one level in the hierarchy, domains would map to data that fits in multi-worker chips, formally referred to as *Compute Node(s)*. Further up one level, domains would map to data that fit in multi-chip nodes (multi-Compute Nodes), and further up in multi-node chassis and cabinets.

Starting from the leaves, each level up the tree would add one hop in the maximum communication distance between any two processing units. Existing petascale systems have a maximum distance of five hops and exascale systems will push this distance to six or seven, and even possibly longer, hops, with a corresponding number of levels in the hierarchical partitioning. This hierarchical partitioning can significantly reduce the communication overhead and mapping algorithm complexity to achieve scalability [1, 4].

The programming model used by the HPC applications should also be considered in the architectural decisions, as well as in the specifications of the runtime system, in order to improve HPC efficiency. Although MPI has been the most popular programming model for developing parallel scientific applications, the PGAS programming model is an attractive alternative for designing applications with irregular communication patterns. It is widely believed that a hybrid flexible MPI+PGAS programming model is an efficient choice for many scientific computing problems and for achieving exascale computing [9]. Figure 11.1 illustrates the proposed ECOSCALE partitioning of future HPC applications, which uses such a hybrid many-core MPI+PGAS programming model. PGAS is used for efficient intra-partition communication where the number of cores is limited by the hop-costs of a specific system instantiation.

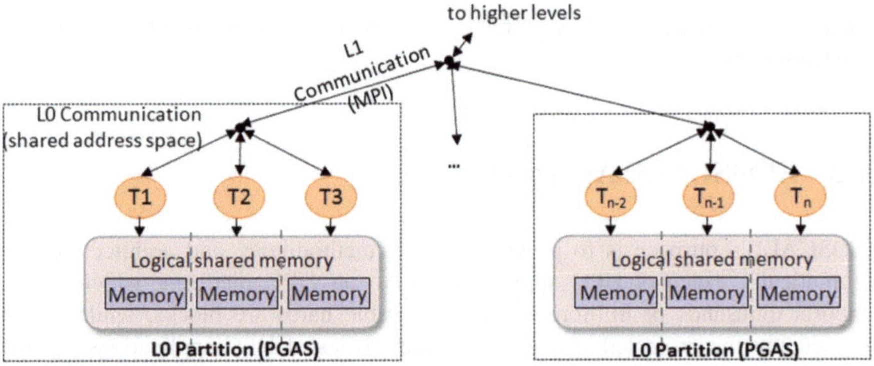

Fig. 11.1 Example hierarchical partitioning (tasks, memory, communication) of an HPC application

Since PGAS and related task scheduling algorithms have important scaling problems, MPI can also be used for efficient inter-PGAS communication.

Exascale performance and energy efficiency are also going to be supported by the extensive use of reconfigurable accelerator technology through the implementation of a novel architecture referred to as *UNILOGIC* (Unified Logic). This is as an extension to the *UNIMEM* architecture proposed and developed within the context of the EUROSERVER project [8]. Specifically, UNIMEM provides a shared partitioned global address space while UNILOGIC extends this concept by providing shared partitioned reconfigurable resources within UNIMEM.

The UNIMEM architecture gives the user the option to move tasks and processes close to data instead of moving data around [8] and, thus, significantly reduces data traffic and the associated energy consumption and communication latency. From the point of view of a processor in a multi-node machine, a memory page can be cacheable at the local coherent node or at a remote coherent node, but not at both. This is the basis of the UNIMEM consistency model, which eliminates global-scope cache coherence protocols, thereby, helping to achieve scalable solutions. Progressive address translation [11] can be further applied on top of UNIMEM in order to provide inter-processor communication.

UNILOGIC enriches UNIMEM by introducing the ability to easily move the acceleration engine to local hardware, for instance, through dynamic partial reconfiguration [12]. The proposed UNILOGIC+UNIMEM architecture partitions the design into several *Compute* nodes that communicate through a hierarchical communication infrastructure, similar to the one shown in Fig. 11.1. These Compute nodes correspond to the partitions of the HPC application. Each Compute node is an entire subsystem including processing units, memory and storage. Within a PGAS domain, consisting of several Workers, the proposed architecture offers (1) a shared global address space that can be partitioned for locality and (2) shared reconfigurable resources that can also remotely access cached data via regular load and store instructions, without using any global cache coherency mechanism to keep a local cache coherent.

11.3 ECOSCALE Approach

ECOSCALE's purpose is to provide a novel methodology and architecture that automatically executes HPC applications onto an HPC platform, which, itself, supports thousands or millions of reconfigurable hardware blocks, while taking into account the projected trends and characteristics of HPC applications. Within this context, ECOSCALE aims at introducing Field-Programmable Gate Array (FPGA)-based acceleration as an integral part of the processing nodes within the UNIMEM+UNILOGIC system architecture and adapting them to work in an HPC environment. Thus, its novel framework provides the locality and scalability model for FPGA-based acceleration from the ground up. In order to efficiently do so, we follow a holistic approach providing solutions for all aspects of an HPC environment, ranging from architecture and runtime management and optimisation, to High-Level Synthesis (HLS) and hardware virtualization.

The proposed hardware (HW) design consists of a stack of three interdependent HW abstraction layers, as shown in Fig. 11.2. At the bottom layer, the proposed hardware architecture provides the basic hardware components and functionality in order to efficiently use the available HW resources, for instance, CPU, memory and reconfigurable hardware.

In the middle layer, a middleware provides the primitives to reconfigure hardware blocks at runtime, while an HLS tool provides the synthesised application tasks to the middleware. Finally, in the top layer, a runtime system schedules tasks inside a PGAS partition, provides the MPI primitives for communication between PGAS partitions and decides at runtime which functions of the accelerated application should be implemented and executed in reconfigurable hardware as well as where data should be placed for locality.

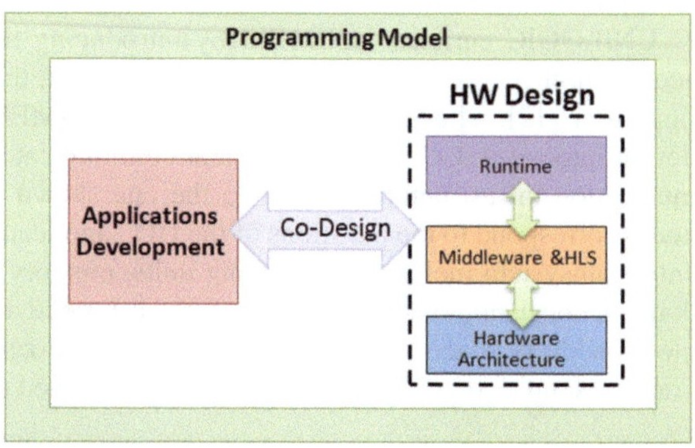

Fig. 11.2 The ECOSCALE framework

11.3.1 The ECOSCALE Architecture

The system architecture uses CPUs, memory and reconfigurable cores (often described as accelerators) in a highly parallel manner. Driven by the characteristics and trends of future HPC applications and following the high-radix partitioning of an HPC application, the proposed UNILOGIC+UNIMEM architecture logically partitions hierarchically the hardware resources, such as CPUs and reconfigurable hardware into several interconnected *Compute Nodes*, corresponding to the PGAS partitions of the application, which are further partitioned into several Workers, depending on the physical structure of the system. Thus, one or more Compute Nodes create an entire and independent PGAS subsystem including several Workers and offer:

1. UNIMEM: a shared space that allows Workers to communicate via regular loads and stores without global cache coherence and
2. UNILOGIC: shared partitioned reconfigurable resources that share the UNIMEM space with software tasks.

Other existing architectures either require a global cache coherent mechanism, which simply cannot scale, or support only Direct Memory Access (DMA) operations, which are not efficient for small data transfers such as messages to synchronise remote threads or to configure a remote peripheral [17]. The UNIMEM architecture allows moving tasks and processes closer to data instead of moving data around [12].

The proposed HPC architecture consists of several Workers communicating through a multilayer interconnection, Fig. 11.3. The actual number of Workers inside a Compute Node depends on the integration capabilities of future technologies. Each Worker is an independent computing unit that can execute, fork and join tasks or threads of an HPC application in parallel with other Workers.

It includes a CPU, a reconfigurable block and an off-chip DRAM memory. The communication and synchronisation between Workers is performed through a mul-

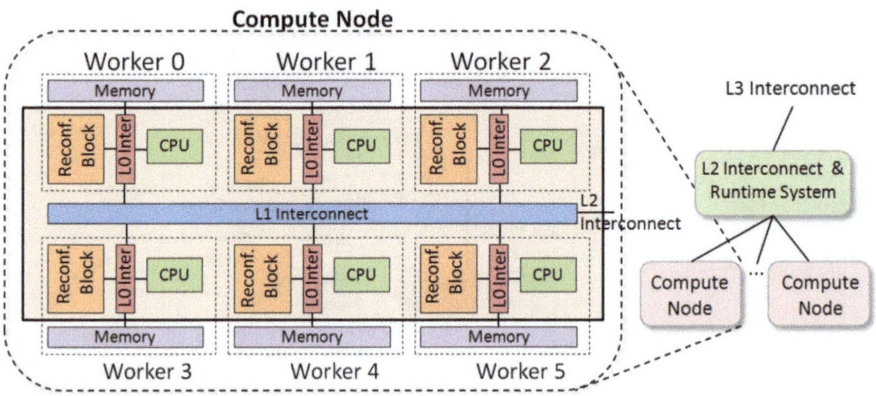

Fig. 11.3 The ECOSCALE compute node hardware architecture

tilayer interconnection, which allows load and store commands, DMA operations, interrupts and synchronisation between Workers within a Compute Node. The Compute Nodes are PGAS subsystems that correspond to the application's PGAS-based partitions shown in Fig. 11.1. Matching the application's logical topology of Fig. 11.1, the Compute Nodes are interconnected through an MPI-based multilayer interconnection.

The communication overhead between a CPU and a hardware accelerator, i.e. the reconfigurable block inside a Worker, is one of the most crucial challenges. A few years ago only explicit memory transfers between the host memory and the accelerator's memory were supported, like in a GP-GPU. Recent technological advances allow the integration of the host CPU and hardware accelerators on the same chip, and thus, hardware accelerators can now access the host memory directly. Such a typical ARM-based system [5] is depicted on the left of Fig. 11.4. However, there are still important limitations that we will tackle in this project.

Subsequently, in the architecture of Fig. 11.4, the ARM Cache Coherent Interconnect supports two types of coherent ports in order to provide hardware coherency in the system: (1) ACE ports, which can be used by masters containing caches, such as a processor, and (2) ACE-lite ports, which can be used by masters that do not have hardware coherent caches. ACE-lite ports are traditionally used for hardware accelerators such as GPUs and FPGAs, as shown in the figure.

Furthermore, Fig. 11.4 includes the block diagram of a single ECOSCALE Worker. This architecture extends the one displayed on the left in the following manner. Accelerator blocks act, according to UNIMEM terminology, as a Unit of Compute, and hence, they can interface directly with any other UNIMEM Unit of Compute where each unit caches its local data coherently. Similarly, each accelerator can cache its local data and provide coherent access from remote UNIMEM units. If a single accelerator block needs to span across multiple FPGA local memories, the FPGA units can then provide their own coherence schemes independent of UNIMEM.

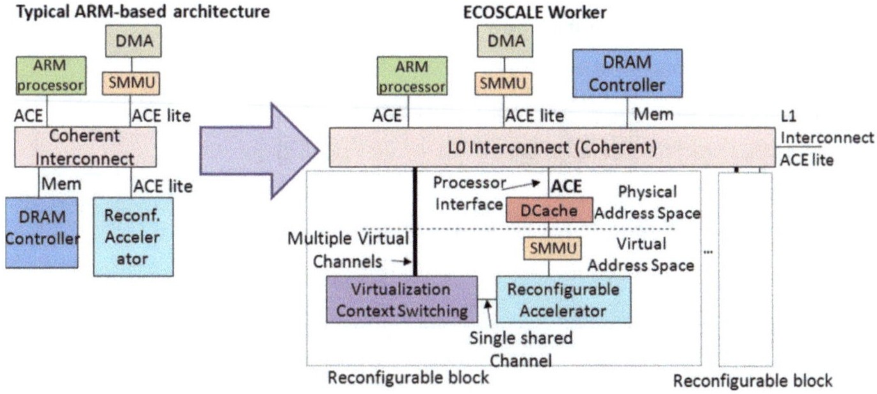

Fig. 11.4 Conventional ARM-based system versus block diagram of ECOSCALE worker

The reconfigurable resources are typically configured to use physical addresses in order to access shared variables. Since only the OS (or the hypervisor in virtualized systems) has access to the physical address space, the intervention of the OS (or the hypervisor) is unavoidable. A dual-stage I/O MMU, such as the ARM SMMU shown in Fig. 11.4, can resolve this problem by translating virtual addresses to physical addresses in hardware. Hence, by using an I/O MMU, the proposed architecture will allow user-level access to the reconfigurable accelerators.

Furthermore, virtualisation and context switching enables multiple tasks or threads of an HPC application to share a single CPU in order to maximise the utilisation of the CPU resources. Similarly, our architecture will support coarse-grain time-sharing of the reconfigurable resources through partial runtime reconfiguration. Moreover, it will support fine-grain sharing of those FPGA resources, where a function implemented in hardware can be called by different tasks or threads of an HPC application in parallel, through the Virtualisation block of Fig. 11.4. The Virtualisation block and the HLS tool provide a mechanism to execute multiple function calls, from different virtual machines, in a fully pipelined fashion.

Sharing of the limited reconfigurable resources between Workers is very important. Thus, within a Compute Node, any Worker can access any reconfigurable block, even remote blocks that belong to other Workers, through the multilayer interconnect of Fig. 11.3. Moreover, the $L0$ Interconnect in this example system provides an external ACE-lite port (connection to $L1$ interconnect in Fig. 11.4) that can be used by remote reconfigurable blocks to make coherent accesses. However, since this is not an ACE port, the remote Reconfigurable block should disable its data cache and would not be as efficient as a local one.

The inter-Worker communication as implemented via UNIMEM is shown in Fig. 11.5. With UNIMEM, every independent ECOSCALE Worker, including those of different Compute Nodes, can potentially have access to the memory physically located anywhere in the system, i.e. the memory that comes as part of a Worker. Not only that, but a memory can be shared among local as well as remote requests, with the latter case shown in Fig. 11.5.

In this particular instance, a remote Worker (Worker 0) is faced with increasing storage demand and uses the memory of Worker 1 that happens to be sitting idle through a PHY to Global and Global to PHY address translation process, thereby, extending the capabilities of the system and avoiding limitations in available memory. Naturally, the address translation process is what the system MMU in the ECOSCALE architecture is responsible for performing.

Subsequently, a very similar concept is applied to the case of UNILOGIC and a simple illustration is shown in Fig. 11.6. As mentioned, UNILOGIC offers remote access to reconfigurable resources, i.e. the reconfigurable hardware of each Worker can become available to any application that may require it provided that it is available. In the example of Fig. 11.6, Worker 0 can access the programmable logic, i.e. Reconfigurable Block, located at Worker 1 through the same process of PHY to Global and Global to PHY address translation, while that particular block can fetch the data required for processing from the memory (DRAM) of the initiator, i.e. Worker 0. This, however, is one type of scenario that the ECOSCALE archi-

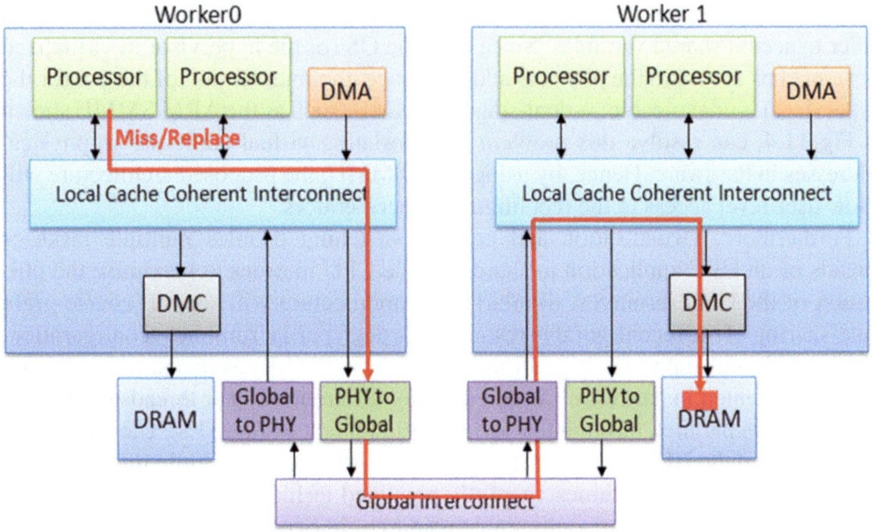

Fig. 11.5 UNIMEM—remote memory sharing

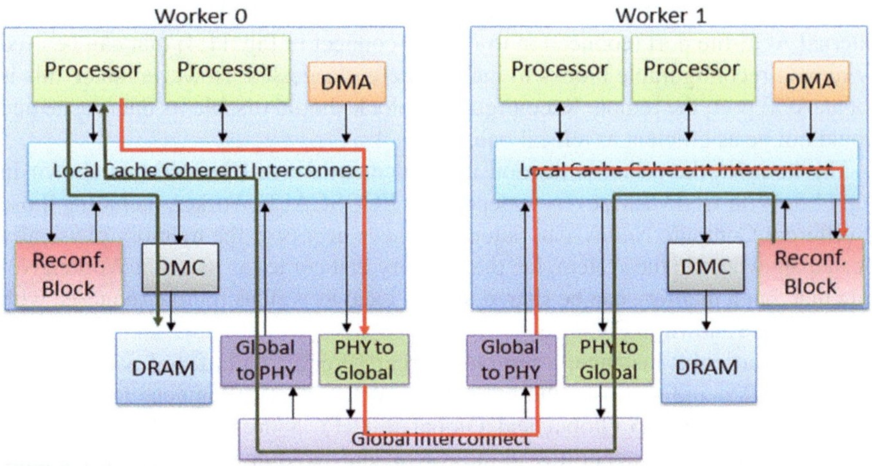

Fig. 11.6 Remote access to reconfigurable resources

tecture shall realise, others will also take place such as a single Reconfigurable Block using data originating from multiple, instead of a single, memory sources or multiple Reconfigurable Blocks using data from a single (or multiple) memory source(s).

11.3.2 Runtime System

The ECOSCALE runtime environment, shown in Fig. 11.7, extends current OpenCL frameworks in three ways. First, by supporting a partitioned global address space within and between ECOSCALE Workers and Compute nodes, via the introduction of new data scoping and consistency abstractions in OpenCL. Second, by extending the semantics and providing a scalable and efficient implementation of OpenCL data transfers between partitions of the address space. This is in addition to data transfers between devices, CPUs and reconfigurable subsystems, within the same address space, by using direct loads and stores from and to remote shared memories. Third, by allowing the programmer to specify functions that can be synthesised in hardware and can be accelerated, on demand, at runtime, depending on the dynamic execution conditions of the system. Furthermore, ECOSCALE implements new algorithms to dynamically partition computation between CPU cores and hardware accelerators on the ECOSCALE nodes.

ECOSCALE has also explored new methods and models for monitoring the execution time complexity and energy consumption of tasks on CPUs and reconfigurable systems, as well as new algorithms for choosing on the fly the most appropriate device to execute each function.

Specifically, input-dependent models are developed that capture the correlation between input/output size, input/output data shape (when available), and data access pattern in memory (model inputs) and execution time and power consumption (model outputs) using one or more CPU cores or accelerator(s). The models are co-designed with the application use cases of ECOSCALE. This effort entails three parts. First, a training part to use the target applications with different realistic inputs, in order to capture static and dynamic properties of the input and record the corresponding execution time and power outputs.

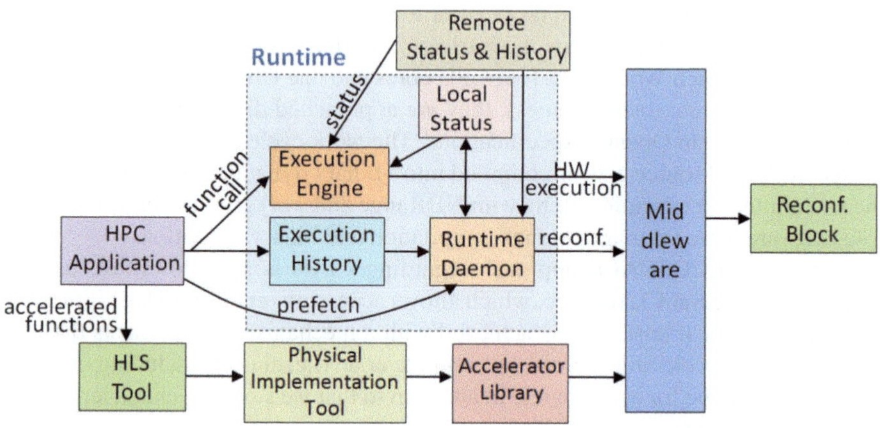

Fig. 11.7 Interaction and control flow between the three abstraction layers

Second, a model building part explores models for predicting outputs from inputs. For this, an array of regression, SVM and PCA techniques are used, building on prior experience on models for predicting execution time and power for the purpose of multidimensional programme adaptation [6]. Finally, actuation part deploys the models with actual running applications, using hardware performance monitors and function instrumentation to capture the static and dynamic properties of the unseen input, and project execution time and power using the trained models with hardwired parameters. This enables the runtime scheduler to judiciously and dynamically select and distribute functions for hardware acceleration.

We also explore methods to minimise non-overlapped communication latency within each ECOSCALE Compute node as well as across nodes in the same cluster. Hence, one scheduler per worker is implemented, to manage the local reconfigurable blocks and the execution of the accelerated functions.

Furthermore, whenever a function is called, a work and data distribution algorithm in the runtime system (included in the Execution Engine of Fig. 11.7) decides whether the function is to be executed in software or in hardware based on the local status and the status of other Workers in the vicinity. To curb the overhead of monitoring remote status, we implement local work queues per worker and infer the status of remote workers via the status of the local queue, using techniques inspired by Lazy Scheduling [18]. A history of the function calls as well as their execution time is stored in a History file (Execution History block). Consequently, the runtime scheduler/daemon reads periodically the system status and the History file in order to decide at runtime what functions should be loaded on the reconfiguration block.

Moreover, there are two partitioning problems that the runtime has to manage. First, the partitioning of data and second the partitioning of the computations. The mechanism for data partitioning is quite straightforward and has been addressed by the use of a library that allows performing remote memory allocations through UNIMEM. Therefore, the job of the runtime is simply to track the current memory allocations and executions and then to allocate the available memory according to the allocation algorithm, which takes all of that into account.

Partitioning computations are more tricky as we have two types of accelerators at our disposal on each Worker, i.e. four CPU cores and one UltraScale+FPGA. Hence, due to their inherent distinctiveness, they are approached differently. CPUs behave normally relative to OpenCL specifications. The work cycle of a typical accelerator is this: the kernel source code is compiled into a binary dynamically and the binary is then submitted for execution along with NDRange and workgroup sizes information. The software driver manages kernel compilation and binary execution.

Hence, the ECOSCALE approach regarding CPUs is as follows. We use the readily available SnuCL runtime, which allows accessing remote CPUs as if they are available on a local host. This is achieved with a mechanism which tracks the load of all the CPU accelerators across the compute node (i.e. cluster) including the work pending in a queue for each device. In addition to that we store the characteristics of previous executions.

Subsequently, the workflow is as follows: the work (the dataset and the kernel) is submitted to a group device. The dataset placement is delayed until the kernel arrives. When the routing logic sees that the kernel has been submitted, it uses the information about the previous runs together with information about the current workload in order to make the routing decision of where to place the dataset and the computations.

Moreover, the routing mechanism is implemented at the NDRange level of granularity. Pending the resolution of a number of problems, the NDRange is decomposed into workgroups and the execution is managed at the workgroup level of granularity.

On the other hand, FPGAs do not map to the CPU model very well. Hence, reconfiguration is achieved through the BitMan API [15]. The computations are started through separate memory mapped APIs. The actual bitstream with accompanying information is specified in the XML file in the proprietary format and, in general, any FPGA can be addressed over UNIMEM as their interface is memory mapped.

In order to pull all these parts together an FPGA Management Layer is utilised, responsible for device discovery, resource management, device reconfiguration, computation submission and execution management. That Management Layer acts as a group device, which combines many physical FPGAs into one.

Hence, the resulting workflow is similar to that of the CPU: the placement of the dataset is delayed until the arrival of the kernel. It is only at that time that a routing decision is made. The difference to the CPU approach is that the routing algorithm for an FPGA needs to take into account the reconfiguration overhead and the bitstream footprint as well as other parameters. Similar to CPU, the NDRange is partitioned into workgroups, which are then scheduled and managed independently.

11.3.3 Middleware and High-Level Synthesis

The middleware bridges the gap between the reconfigurable hardware and the software parts of the full application, providing the means to enable a fully software-driven development flow. The middleware plays two main roles, i.e. provides the partial-reconfiguration toolset and the SW-HW communication library. Thus, first, it performs partial reconfiguration at runtime. This includes the development of a low-level driver back end that adds virtualisation features, such as de-fragmenting the reconfigurable resources, accelerator migration and pre-emptive hardware execution. Second, it provides a communication library and API in order to call any function that is implemented in hardware.

ECOSCALE supports hardware-assisted virtualisation, Fig. 11.8, in order to increase the performance and to lower the power consumption. Also, it extends the HLS tool developed in the FASTCUDA project [14] and holds the main focus on effectively exploiting the huge cost/performance trade-off space provided by ECOSCALE, while requiring minimal intervention and no specific hardware design experience from the programmer.

The ECOSCALE HLS tool provides a way to specify performance and area constraints, and then automatically explores high-performance hardware implementation techniques, such as pipelining, loop unrolling, as well as data storage and data-path

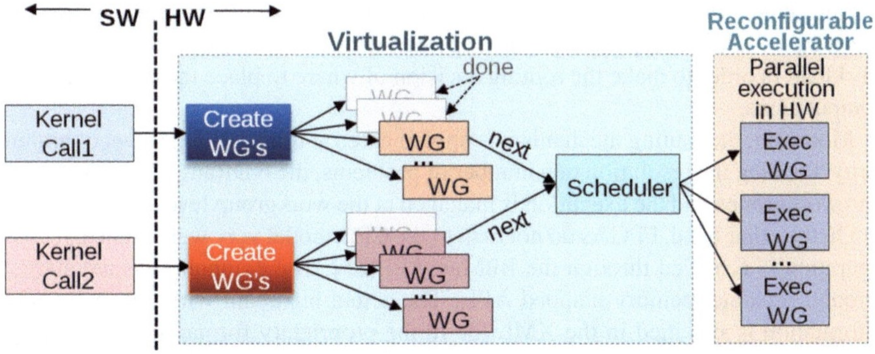

Fig. 11.8 Virtualisation block accepts several calls of the same kernel and schedules their work-groups to the reconfigurable accelerator

partitioning and duplication, starting from a non-hardware specific OpenCL model. Current HLS tools require an experienced designer to take architectural decisions, such as the DRAM port parallelism, the local data memory partitioning and so on. These are automated considerably while retaining designer control, if and when needed.

The tool generates, at compile time, a library with the hardware implementations of those functions that will be implemented on reconfigurable resources. These implementations are transformed with the help of a physical implementation tool, which extends the existing GoAhead framework [3], automatically into an accelerator module library. This includes the steps of resource budgeting, floorplanning, communication infrastructure synthesis and physical constraint generation for the reconfigurable fabric, place and route tools, as well as the final partial bitstream assembly. By minimising module bounding boxes and by using configuration data compression [13], memory requirements are reduced as well as configuration latency and configuration power consumption.

At runtime, the system can use this library in a very flexible manner. For example, we consider chaining together different accelerator modules for building longer complex processing pipelines, when needed. This increases substantially the amount of processing that is carried out per unit of transferred data and results in significant energy savings.

Finally, Fig. 11.8 shows how multiple function calls of the same function is executed in parallel. Each function call in our system is essentially an OpenCL kernel call, which can be mapped onto multiple work items. Upon reception of a kernel call, the Virtualisation block starts configuring and scheduling the synthesised hardware onto the reconfigurable accelerator. For each kernel call, the Virtualisation block needs to remember only what workgroups have been executed or scheduled for execution so far. Moreover, it can mix the execution of the kernel originating from different function calls, as well as provide Quality of Service (QoS) by controlling the rate at which the workgroups are executed. The Accelerator includes a number of the same workgroup implementations, which corresponds to the number of the outstanding workgroups that can be scheduled in parallel by the Virtualisation block.

11.3.4 Programming Model

ECOSCALE develops co-designed HPC applications based on hierarchical data partitioning to achieve locality and reduced data traffic and associated power consumption and latency, and the focus of the programming model efforts is in two directions. The first is to extend OpenCL to support multiple Workers, distributed command queues and transparent command queue management across Workers in a Compute node. The second direction is towards providing a transparent substrate to achieve data access locality from OpenCL instances running in Workers. The global memory in each Compute node is treated as a collection of NUMA domains accessible via the UNIMEM interface. Consequently, topology-aware global memory allocators are used by the OpenCL runtime for implicit data allocation, migration and replication between workers.

Traditionally, an OpenCL programming model would represent each physical accelerator, i.e. GPU or FPGA, as an object of type `cl_device_id`. The dataset would be submitted to that chosen device along with a kernel while the kernel represents one unit of execution. Upon kernel submission, the user instructs the runtime how many kernels to instantiate, i.e. the so-called NDRange, and also how to group them. An instance of a kernel is known as a work-item and a group of work items is referred to as a workgroup. Any work-item can access any part of the dataset. The work items are synchronised within a group through the use of barriers, but are not synchronised across the workgroups.

Whereas, originally, the OpenCL programming model was intended for execution on just one host, numerous projects such as SnuCL, VCL and VOCL [2, 10, 20] have extended that model by making remote accelerators accessible as if they are located all on one host.

Thus, as a further extension to that model, ECOSCALE introduces the notion of a group device. Contrary to an ordinary OpenCL device, the group device corresponds not just to one physical device, but to many devices (including remote devices, i.e. on other Workers). By using the group device, the developer no longer has to partition the data and computations as this is done automatically within the routing logic of a group device. The partition patterns can be specified by the user, and these are the REPLICATE, BLOCK, CYCLIC and HALO [19].

11.4 Conclusions

Today's technologies and architectures cannot efficiently scale to exascale. A holistic approach tailored to the characteristics and trends of future HPC application is required. Towards this end, ECOSCALE employs a novel hardware architecture, runtime system and programming model in order to be able to directly map the HPC application's hierarchical structures onto hardware resources, while in parallel it takes full advantage of energy-efficient reconfigurable computing. ECOSCALE provides solutions for all the aspects of an HPC environment, ranging from architecture and runtime optimisations, to partial reconfiguration, HLS and hardware virtualisation.

Finally, the ECOSCALE framework introduces two novel technologies, i.e. UNIMEM and UNILOGIC. These technologies are major contributors in the concept of resource sharing independent of whether they are remote or local and span from those of memory to those of processing through reconfigurable logic.

Acknowledgements This research project is supported by the European Commission under the H2020 Programme and the ECOSCALE project (grant agreement 671632).

References

1. Abdel-Gawad A, Thottethodi M, Bhatele A (2014) RAHTM: routing-algorithm aware hierarchical task mapping, SC
2. Barak A, Ben-Nun T, Levy E, Shiloh A (2010) A package for OpenCL based heterogeneous computing on clusters with many GPU devices. In: IEEE international conference on cluster computing workshops and posters (CLUSTER WORKSHOPS), pp 1–7. https://doi.org/10.1109/CLUSTERWKSP.2010.5613086
3. Beckhoff C, Koch D, Torresen J (2012) GoAhead: a partial reconfiguration framework. In: FCCM
4. Chung I-H, Lee C-R, Zhou J, Chung Y-C (2011) Hierarchical mapping for HPC applications. Parallel Process Lett
5. CoreLink CCI-400 Cache Coherent Interconnect. http://www.arm.com/products/system-ip/interconnect/corelink-cci-400.php
6. Curtis-Maury M, Shah A, Blagojevic F, Nikolopoulos D, de Supinski B, Schulz M (2008) Prediction models for multi-dimensional power-performance optimization on many cores. In: PACT, pp 250–259
7. Dongarra J, Beckman P, Moore T et al (2011) The international exascale software project roadmap. IJHPCA 25(1):3–60
8. Durand Y et al (2014) EUROSERVER: energy efficient node for european micro-servers. In: Euromicro DSD
9. Jose J, Potluri S, Subramoni H, Lu X et al (2014) Designing scalable out-of-core sorting with hybrid MPI + PGAS. In: 8th PGAS programming models
10. Kim J, Seo S, Lee J, Nah J, Jo G, Lee J (2012) SnuCL: an OpenCL framework for heterogeneous CPU/GPU clusters. In: Proceedings of the 26th ACM international conference on supercomputing, ICS '12, 2012, San Servolo Island, Venice, Italy. ACM, New York, NY, USA, pp 341–352. https://doi.org/10.1145/2304576.2304623. ISBN 978-1-4503-1316-2
11. Katevenis M (2007) Interprocessor communication seen as load-store instruction generalization. In: Bertels K et al (eds). Delft
12. Koch D (2012) Partial reconfiguration on FPGAs—architectures, tools and application. Springer
13. Koch D, Beckhoff C, Teich J (2009) Hardware decompression techniques for FPGA-based Embedded Systems, ACM TRETS
14. Mavroidis I, Papaefstathiou I, Lavagno L et al (2012) FASTCUDA: open source fpga accelerator & hardware-software codesign toolset for CUDA kernels. In: DSD

15. Pham KD, Horta E, Koch D (2017) BitMan: a tool and API for FPGA bitstream manipulations. In: Design, automation and test in Europe conference exhibition (DATE), pp 894–897
16. Prisacari B, Rodriguez G, Heidelberger P, Chen D, Minkenberg C, Hoefler T (2014) Efficient task placement and routing of nearest neighbor exchanges in dragonfly networks. In: 23rd HPDC. ACM
17. Showerman M et al (2009) QP: a heterogeneous multi-accelerator cluster. In: High-performance clustered computing
18. Tzannes A, Caragea GC, Vishkin U et al (2014) Lazy scheduling: a runtime adaptive scheduler for declarative parallelism. ACM Trans Program Lang Syst
19. Yan Y, Lin P, Liao C, de Supinski B, Quinlan D (2015) Supporting multiple accelerators in high-level programming models. In: Proceedings of the sixth international workshop on programming models and applications for multicores and manycores, PMAM '15. ACM, New York, NY, USA, pp 170–180. https://doi.org/10.1145/2712386.2712405
20. Xiao S, Balaji P, Dinan J, Zhu Q, Thakur R, Coghlan S, Lin H, Wen G, Hong J, Feng W (2012) Transparent accelerator migration in a virtualized GPU environment. In: Proceedings of the 2012 12th IEEE/ACM international symposium on cluster, cloud and grid computing (CCGRID 2012). IEEE Computer Society, Washington, DC, USA, pp 124–131. https://doi.org/10.1109/CCGrid.2012.26
21. Website of ECOSCALE. http://www.ecoscale.eu/

Chapter 12
On Optimizing the Energy Consumption of Urban Data Centers

Artemis C. Voulkidis, Terpsichori Helen Velivassaki and Theodore Zahariadis

12.1 Introduction

Data Centers (DCs) are responsible for the absorption of an enormous and steadily increasing amount of energy, bearing considerable impact on the environmental conditions worldwide, particularly in terms of physical resources depletion and CO_2 emissions. In 2011, the total energy consumption of DCs was estimated at around 271 billion kWh, enough to power up all residential households of industrialized countries such as France or the UK or, equally, comparable to the total amount of energy consumed by Italy [1], approximately 7% of the US total energy consumption [2]. Just the Microsoft DC in Washington (Quincy) consumes 48 mW, which is enough to power almost 40,000 households [3, 4]. In the same context, the energy consumption of DCs greatly affects the global economy; modern DCs may have operational costs as high as $5.6 M [5] per year, while in 2010 and 2011, USA spent approximately $35 billion in serving DC power needs.

Besides electricity consumed for supporting the computing services provided by the DCs, a huge amount of energy is also consumed for the cooling of computing servers. To lower this waste of energy, DC containment strategies (both hot aisle and cold aisle) are widely regarded as the starting point for energy efficiency best practices. Moreover, the so-called "Green DCs" aim to use green energy sources (e.g., photovoltaic cells, geothermal power, hydroelectric energy, etc.) for normal operations and cooling purposes. The results are, in many cases, impressive, but they

A. C. Voulkidis (✉) · T. Zahariadis
Synelixis, Chalkida, Greece
e-mail: voulkidis@synelixis.com

T. Zahariadis
e-mail: zahariad@synelixis.com

T. H. Velivassaki
Singular Logic, Kifisia, Greece
e-mail: tvelivassaki@ep.singularlogic.eu

C. Kachris et al. (eds.), *Hardware Accelerators in Data Centers*,
https://doi.org/10.1007/978-3-319-92792-3_12

still represent a minority of the deployed DCs and even in those cases, the intermittent nature of green energy sources makes the need for their effective integration to the energy network and for stable smart grid operation more actual than ever.

In this framework, recent advances in the digitization of the energy sector through the emergence of the notion of the smart grid introduced significant challenges as to:

(a) how to support the proliferation of urban DCs, namely DCs located in or in the close vicinity of urban environments such as cities of any size, with the subsequent increase in the energy demand, further augmented by the relevant suboptimal energy management. Generally, the average utilization in terms of computational resources in DCs is relatively low, often not reaching 30% of the maximum server load capacity and only 10% in case of facilities that provide interactive services [6]. The operational security-driven over-dimensioning of DCs and the increased number of underutilized servers have significantly increased the energy consumption, leading to increasing energy needs for cooling DC IT equipment. Towards energy saving, DC containment strategies (both hot and cold aisle) are widely regarded as the starting point for energy efficiency.
(b) how to manage the operational instability of the smart grids in an increasing energy demand context and alleviate their difficulty to follow the electricity demand-response model. As energy generation globally shifts away from fossil fuels in favor of renewables, smart electricity grids are becoming increasingly harder to operate and maintain.

In this direction, several research activities focus on energy-efficient DC operation and many energy reducing methods have been proposed. Particularly, heating, ventilation, and air conditioning control (HVAC) techniques exploit the nonlinear dependence of the energy consumption on the DC air temperature. On the other hand, numerous techniques trade-off performance for energy saving, including CPU scaling/voltage control (Dynamic Voltage/Frequency Scaling—DVFS), load migration and load shifting [7]. To elaborate, controlling CPU voltage and scaling its operating frequency results in reducing its performance, simultaneously baring significant energy savings, whereas load migration and load shifting refer to the procedures of moving the computing loads in the spatial and temporal domain, respectively. Specifically, load migration consists of transferring the computing requirement of a service from one server host to another, whereas load shifting postpones serving a load at later time instances, e.g., when the energy required for this service is cheaper or abundant.

In [8], a survey on the most popular techniques for improving the energy efficiency of large scale data centers is presented. The authors classify the existing literature into three different optimization levels, namely optimization at node (host), optimization of infrastructure, and optimization of virtualized environments. These optimization levels include, among others, DVFS, software and hardware improvements, task scheduling and thermal management and virtual machines (VM) migration. A similar survey has been conducted in [7] where asymmetry-aware scheduling is also surveyed as a way of maximizing performance through proper exploitation of the multi-core capabilities of modern CPUs.

Weiser et al. [9] were among the first to notice that it is more energy efficient to serve a computing intensive, Quality of Service (QoS) constrained application in a just-in-time manner operating at lower CPU frequencies, rather than serving at full processor speed, completing it before its deadline and then idling, introducing the term of DVFS. Since then, numerous approaches have been proposed to optimize DVFS patterns when applied to control CPU [10–13] and RAM [14] (or even both [15]) of physical servers. Specifically, in [10], it is argued that per core DVFS can achieve up to 20% energy savings, a result affirmed by [11] as well, while [12] concludes that although recent advances in transistor lithography technology significantly diminishes the merits of DVFS (due to the reduced dynamic power range), non-negligible power savings may be attained via CPU DVFS. Two energy-aware heuristics taking into consideration different power consumption metrics are presented in [13], indicating that smart execution scheduling assisted by DVFS can greatly reduce server energy dissipation. The authors of [14] explore the trade-offs between performance and energy efficiency when performing DVFS on RAM DIMMs, using linear approximations to model the reduction of performance due to the scaling applied. The experiments conducted indicate that significant energy savings might be achieved by applying DVFS, especially in cases of low memory utilization.

Controlling the network equipment has also been an area of active research and experimentation since, in this manner, significant energy savings may be attained [16]. Indicatively, in [17], the need for energy-aware networking interfaces in large-scale networks is highlighted, whereas [18] evaluates the energy efficiency of networking equipment. Heller et al. [19] presented a hierarchical, tree-based architecture that can significantly reduce the energy consumed by networking devices in large DCs by turning off idle equipment. Active network equipment switching is also the target of the optimization methods proposed in [20].

The optimal allocation of the computing load in DC context has also attracted the attention of the research community during the last years [21–24]. In [21], load balancing in very large-scale DCs running network intensive applications (Windows Live Messenger services) is examined. The authors claim that proper estimation of future load may lead to increased energy savings, without compromising end-users Quality of Experience (QoE). Voorsluys et al. elaborate on the cost of live VM migration, concluding that under heavily used DCs live migration might result in non-negligible delays and, even, QoS breakage. However, under normal circumstances it should engender barely noticeable changes to end-users QoE. A low-complexity variation of the well-known Best Fit Decreasing (BFD) algorithm to optimally allocate VMs in a DC is proposed in [23]. This algorithm prioritizes the servers that are going to accommodate the migrated load according to the overhead power that they will need to properly serve this load.

A load migration scheme to maximize the use of green energy through proactive scheduling is proposed in [25]. Specifically, the authors propose a load migration scheme on a network of spatially dispersed DCs, each one supported by one or more Renewable Energy Sources (RES), based on predictions related to the short-term future DC load and the availability of green power near the DCs. The corresponding simulation results suggest that a large amount (up to 30%) of brown energy may be replaced by green energy, at the cost of slight (approximately 2%) increase of energy consumption.

Based on the related work and the available methods for reducing energy cost, the present paper classifies the available degrees of freedom outlining the general framework of energy-efficient DC operation. In this course, a general architectural approach is described comprising an ecosystem of collaborative DCs, exploiting the available degrees of freedom and the emerging levels of optimization. The proposed architecture facilitates the implementation of four different optimization levels and has been developed in the context of the European Union project: "Data centres Optimization for energy-efficient and enviromentalLy Friendly INternet (DOLFIN)" [26]. A similar approach is followed by another EU project "Green nEtworked data centers as energY proSumers in smaRt city environments" (GEYSER) [27]. The next paragraphs are the result of collaborative liaison research developments of these two projects, with a focus on DOLFIN for reasons of brevity. Having defined a flexible system architecture, the present paper examines the leeway provided by software-defined networking (SDN) to achieve DC systemic flexibility. Specifically, system-defined infrastructure (SDI) is employed to achieve efficient VM scheduling, server resources reclamation, and substantial energy savings during periods of low server utilization. Since the average server utilization in a DC is often well below 30% of the maximum server load [28, 29] or even 10% in case of DCs providing interactive services [30], an efficient VM scheduling scheme reducing energy cost during low server utilization is essential. In this course, the proposed scheme focuses on allocation of VMs accommodating low load, solely based on SDN. As a result, newly admitted VMs are placed on servers in accordance with a policy minimizing reserved physical resources (i.e., memory) and the implicit operating energy cost. Thus, the energy cost incurred by migration, relocation, and other load optimization techniques is discounted, during low server utilization. However, in cases of high load, the proposed scheme must be used complementary with load optimization techniques and this remains to be tackled by future works within the scope of the two aforementioned EU projects.

The remainder of the chapter is organized as follows. Section 12.2 presents the available degrees of freedom and the emerging levels of optimization for energy-efficient DC operation. Section 12.3 provides the flexible architecture design of DOLFIN facilitating the implementation of the optimization process described in Sect. 12.2. Next, Sect. 12.4 formulates the VM scheduling problem, proposing a VM allocation scheme based on the "First Fit Decreasing" algorithm (FFD) [31]. Section 12.5 presents the evaluation of the overall DOLFIN solution and, finally, Sect. 12.6 concludes the paper and presents relevant perspectives.

12.2 DC Energy Efficiency Characteristics

The techniques described in the introduction pave the way for the implementation of an elastic cloud able to manage cloud resources and handle client requests in a flexible way, through the efficient scheduling of VMs accommodating the cloud load. This allows for energy-efficient DC operation, which is essential from an operational point of view, especially given the large scale of modern DCs. Indicatively, the average DC energy consumption was equal to that of 25,000 households in 2007 and doubles every five years [24]. Hence, the energy-efficient and flexible DC operation is of essence, whereas DC interaction with the energy network can be exploited to counteract the adverse effect of the stochastic nature of renewable energy production, stabilizing smart grid operation by dynamically allocating the DC resources. The high-level architecture of such a federated DC network interacting with the underlying energy network is presented in Fig. 12.1.

In this course, the available techniques enabling seamless operation of such a flexible federated DC network need to be featured and exploited. Based on the literature review presented in the previous paragraphs, these techniques include, among others, DVFS, load migration, load relocation, load shifting, HVAC control, and Service Level Agreements (SLAs) renegotiation. SLA Renegotiation refers to the process of automatically negotiating and agreeing with DC customers in lowering the SLA performance parameters in favor of decreased service prices and minified energy consumption. This process can be automated, based on customer profile information and a predefined customer latitude.

In order to exploit the above degrees of freedom in the flexible federated network of Fig. 12.1, each technique needs to be assigned to the respective network level amenable to the specific optimization, namely the servers' rack level, the DC

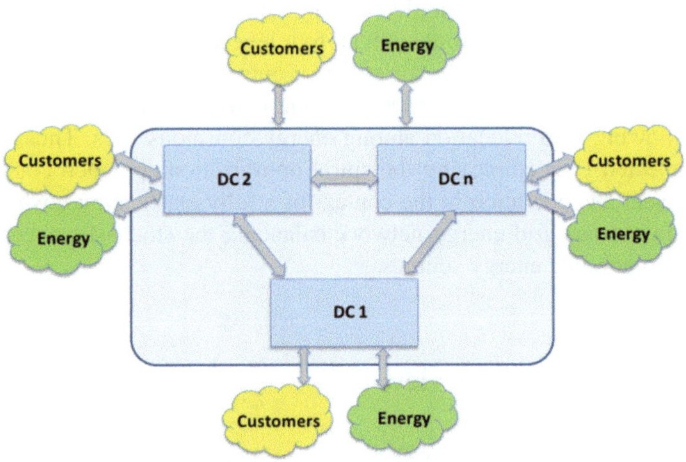

Fig. 12.1 Federated DC network interacting with the energy network

segment level (e.g., a server room), the DC level, and the Federated DC Level. In particular, DVFS takes advantage of the quadratic dependence of CPU power on CPU supply voltage and the linear dependence on clock frequency to trade off performance in favor of substantial CPU power savings. And since CPU is the major energy-consuming part of the servers' components, DVFS has a great impact on server power, thus being amenable to Server's Rack Level optimization.

Load migration refers to the process of moving VMs accommodating computing load across the same DC, to consolidate loads in the same server or DC segment, thus allowing the shutdown of free servers providing substantial energy savings. Therefore, load migration implicitly refers to the DC Segment and the DC Level along with HVAC. HVAC refers to the heating, ventilation, and air conditioning control of the DC or DC segment to trade off performance in favor of energy savings, exploiting the nonlinear relation of energy savings and temperature decrease.

Load relocation describes the process of moving VMs across different DCs, exploiting a lower energy price at a synergetic—interconnected DC or a possible energy surplus provided by a renewable energy source near the synergetic DC. Load relocation is applicable to the federated DCs level, distinguishing itself from load migration since moving of VMs between DCs requires moving of the swap space and disk space introducing a significant stress to the system and network following the movement of a substantial amount of data.

Load shifting defines the moving of VMs in the time domain, i.e., the postponement of a certain load in favor of energy or cost savings to time periods when energy prices are lower or non-negligible energy surplus is available. Load shifting should always be in line with the contractual customer SLAs and, just like SLA renegotiation, is applicable to all levels of optimization. That is, since the load of a server rack, a DC, and a DC network can all be postponed, provided the customer has consented to the load shift and the customers of all levels can renegotiate the existing SLAs.

The above classification of the system degrees of freedom gave rise to four different optimization levels that should be dealt with. In this course, we a distributed spiral optimization process dealing with all four levels of optimization. Specifically, during a single optimization cycle, the energy is optimized first by an internal control loop at servers' rack level, next at DC segment or DC level and then at federated DC level allowing load relocation among energy-conscious DCs. Thus, a network of interconnected DCs employing the spiral optimization approach could provide energy-efficient DC operation in the context of a fully elastic cloud, while playing a key role in a smart grid energy network balancing the stochastic energy surplus provided by renewable energy sources.

12.3 Architectural Overview

Although both DOLFIN and GEYSER projects share common characteristics and scope of research, their approach is fundamentally different, DOLFIN emphasizing on the DC level optimization and GEYSER embracing the smart city ecosystem by means of an energy marketplace enabling green energy trading among DC and smart city actors. The analysis following in the next paragraphs is derived mainly on the DOLFIN approach.

Following the classification of all available degrees of freedom, a flexible architecture has been specified to facilitate the implementation of the DOLFIN optimization process. Such an architecture consists of three key components, namely an Energy Consumption Optimization Platform (eCOP), an energy-conscious Synergetic Data Centers (SDC) module, and an Energy Broker facilitating the interaction of the DOLFIN entity with the underlying energy network.

eCOP lies in the core of DOLFIN platform and is responsible for the energy consumption optimization at DC level. In this sense, eCOP is responsible for implementing the lower three of the optimization levels, employing all techniques presented earlier, except for load relocation. On the other hand, SDC is responsible for applying the fourth level of optimization, employing load relocation along with universal load shift and SLA renegotiation, at DOLFIN level. Finally, the Energy Broker is the liaison of the DOLFIN ecosystem with the energy network, providing insights related to the variable smart grid energy demands, driving DOLFIN operation toward an operational status of balancing the smart grid network.

The above architecture can be further analyzed focusing on subcomponents of eCOP and SDC, facilitating existing and new DC functionalities. DC components available in contemporary DCs that are also essential in the context of the proposed flexible architecture are: a DC Operator (DCO) Hypervisor Manager, a DCO Appliance Manager and a DCO Monitor/Collector. The DCO Hypervisor Manager is responsible for translating high-level decisions into low-level actions and operations, pertaining to ICT hardware management and VM configuration. A DCO Appliance Manager, being the counterpart of the DCO Hypervisor Manager on non-ICT infrastructure, executes high-level decisions related to HVAC and lighting facilities control. Finally, a DCO Monitor/Collector collects operational and energy-related information of both the ICT and non-ICT infrastructure and persists them into the eCOP Data Base.

Apart from the existing DC functionalities, additional functionalities have been defined within the framework of DOLFIN. Specifically, the eCOP Monitor Data Base stores all real-time and historical energy related data of the DC in hand. These data provided by the DCO Monitor/Collector are used along with data collected from all DC components to assist high-level decisions for the achievement of energy-efficient operation. An ICT Performance and Energy Supervisor is also a requisite feature of the proposed architecture, aggregating in a systematic way performance utilization data collected by the eCOP Monitor Data Base, e.g., CPU utilization, memory consumption, etc., of the hosted VMs, to perform an in-depth analysis of the utilization levels of each device. These data are subsequently fed to an Energy-Efficient Policy Maker and Actuator, that is the most intelligent component of the eCOP, to take decisions oriented toward energy-efficient operation, based on the utilization levels of each DC devise, realizing the lower three levels of optimization under consideration.

Furthermore, several new DC functionalities need to be defined to support the network of SDC and realize the fourth level of optimization. At DOLFIN level, a Cross-DC Monitoring Data Collector should be initially employed to collect information from the network routers, to ensure load relocation is feasible from a network point of view given the current traffic and network utilization. Then, a Cross-DC Workload Orchestrator, i.e., a distributed software element of intelligence, gets the SDC decisions regarding load relocation, load shift and SLA renegotiation. The Cross-DC Workload Orchestrator collects information from the Cross-DC Monitoring Data Collector and the DOLFIN Information Data Base to implement the necessary courses of action toward energy optimization. The DOLFIN Information Data Base contains information regarding all DOLFIN architecture components and information regarding the local energy requirements of each DC, i.e., DC energy demand and energy offer from the local smart grid operator. Subsequently, based on the information obtained by the Cross-DC Monitoring Data Collector and the DOLFIN Information Data Base, the Cross-DC Workload Orchestrator realizes the fourth level of optimization implementing the load relocation through the employment of the Cross-DC VM manager. A Cross-DC Manager realizes the interface of interconnected DCs, performing the actual VM relocation cross DCs.

The above-described components collaborate closely to provide flexible and reliable cross-DC communication in the direction of efficient load relocation. However, toward the SLA renegotiation and the energy network interaction, additional components are required. In particular, a SLA renegotiation controller is responsible for

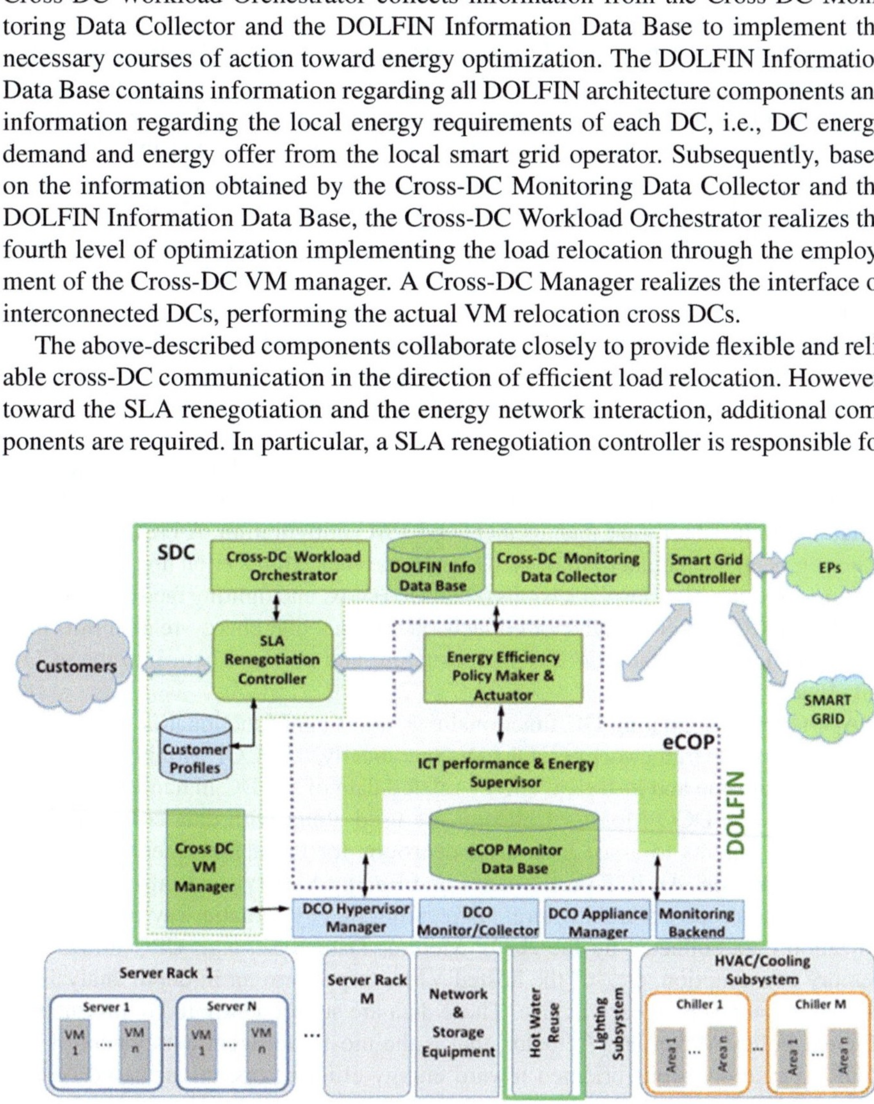

Fig. 12.2 Detailed DOLFIN functional architecture

exploiting the leeway provided by customer agreements trading off performance to energy savings. Specifically, in cases where temporal performance degradation is essential to the DC operation, the SLA renegotiation controller explores the latitude provided by the customer SLAs to provide services of lower performance at a lower price, in favor of energy savings, possibly critical to the DC operation. This task is performed through the notification of the Energy Efficiency Policy Maker and Actuator of the corresponding DC.

Finally, as part of the Energy Broker, a smart grid Controller is required providing information on the local energy requirements of a smart grid and the offered energy prices. This information is collected by the DOLFIN Information Data Base and analyzed by the corresponding local Energy Efficiency Policy Maker and Actuator who acts accordingly.

The above-detailed components and their interaction are presented in a schematic way in Fig. 12.2.

12.4 DC Resource Allocation Methodology

Having elaborated on the flexible DOLFIN architecture and the supported levels of optimization, it is evident that such a DC ecosystem can provide substantial energy savings while underpinning smart grid networks through balancing the stochastic renewable energy. In this course, a network of interworking DCs must be able to exploit the abundant energy, emerging in the context of a smart city through the efficient load migration and relocation to the regions of interest while the energy surplus is available. This imposes the formulation of efficient optimization algorithms for the migration and relocation of loads and remains to be tackled in future works within the framework of DOLFIN. However, prior to this optimization, the present paper explores the prospect of employing SDI to deal with the initial allocation of VMs to servers and the handling of new admission requests.

The benefit from focusing on the SDN aspects of the problem is twofold. First, a consolidated allocation of VMs to servers based on efficient SDN ensures that during low server utilization, which is usually the case [28–30], substantial energy savings can be achieved, without imposing the migration of loads and the energy cost incurred by data movement. Second, during heavy load and high server utilization when load migration can reduce energy consumption significantly, heavy load migration can also lead to delay and Quality of Service (QoS) breakage [22]. Hence, the employment of efficient SDN and the consolidated allocation of VMs prior to the load migration reduces the number of necessary migrations, providing energy-efficient DC operation while meeting the QoS objectives.

In order to efficiently employ SDI, cloud managers, such as OpenStack [32], OpenNebula [33] and Eucalyptus [34] can be used to schedule the allocation of incoming VMs with the objective of minimizing the reserved physical resources and the implicit operating cost. Given the VM reservation of virtual CPU and the pervasive employment of shared storage by different VMs, the allocation of server

CPU and storage to different VMs can be neglected, whereas the only physical resource allocated in a stringent way is the server memory. That is, since prior to the instantiation of a VM physical memory is explicitly reserved. The choice of physical memory has been made in the framework of virtualization services provisioning with shared CPUs and dedicated RAM resources.

In this context, problem of VM scheduling can therefore be considered as a "bin packing" problem [31], where given a finite set $U = \{u_1, u_2, \ldots, u_n\}$ of "items" (i.e., VMs) and a rational "size" (i.e., memory) $s(u)\ \forall u \in U$, a partition of U into disjoint subsets $U_1, U_2, \ldots, U_k$ must be found such that the sum of the sizes of the items in each subset U_i is no more than a respective "bin size" (i.e., server memory) S_i and such that k is as small as possible. Thus, VMs of memory s need to be allotted to servers of memory S, while reserving the minimum number of servers.

Furthermore, the above "bin packing" problem needs to be extended, as the above formulation is tantamount to the problem of allocating newly admitted VMs to free servers. However, the problem needs to account for the allocation of VMs to occupied servers, accommodating previously instantiated VMs, whereas the server memory reserved by completed VMs needs to be reallocated dynamically. In this course, the problem needs to be divided in two: first, VMs need to be efficiently allotted to free servers and second, newly admitted VMs need to be allotted amidst the existing VM allocation, exploiting unreserved memory resources in the direction of a consolidated memory allocation. For both cases of memory allocation, a memory granularity of 512 MB can be assumed which is a typical value encountered in practice.

Following the above rationale, the energy-efficient VM allocation into free servers is equivalent to the "bin packing" problem, hence, the FFD algorithm [31], which constitutes one of the best approximation algorithms for the "bin packing" problem, can be employed, in the direction of a consolidated memory allocation. In this course, the DC servers are indexed based on their energy efficiency, with energy-efficient servers being assigned a lower index. Subsequently, "items" (i.e., VMs) are placed into "bins" (i.e., servers) in order of increasing index. As a result, energy-efficient servers are assigned a higher priority and for instance servers of a Green Room are reserved first, or servers of the same DC segment are reserved prior to remote DC servers to allow remote DC servers to hibernate, providing substantial energy savings.

To consolidate the VM reserved memory and minimize the number of active servers, one could observe that the worst performance of the above algorithm occurs when smaller "items" appear before the larger "items" in the ordering used by the algorithm, since smaller "items" which could be used to fill a half-empty "bin" tend to occupy whole "bins". Hence, the FFD algorithm states that instead of merely taking the items from U in the given order, we first sort them by size and re-index them so that $s(u_1) \geq s(u_2) \geq \cdots \geq s(u_n)$. "Items" are then placed in order of increasing index into the lower indexed "bin" they fit.

Moving to the second part of the problem, that of placing incoming VMs amidst an existing VM allocation, an FFD policy is also assumed to provide a consolidated VM allocation to the extent possible, without the employment of any load migration scheme. Thus, the energy cost emerging from data movement is discounted, whereas loads insusceptible to migration (e.g., live VMs subjected to strict SLA) do not hinder the energy-efficient operation of the DC. However, since load migration is a vital feature of an interconnected DC architecture, the proposed scheme can be consecutively integrated to a load optimization module, providing significant energy savings.

The proposed FFD algorithm sorts every incoming VM and all currently hosted VMs by size. Subsequently, VMs of order index i are placed on servers of greater equal index than the server hosting VM of order index $i - 1$. Thus, the incoming VMs are not placed in the first server of sufficient memory resources, but follow a FFD policy even if this leads to an unconsolidated memory allocation. If the strictly consolidated memory allocation is a requisite feature, a load migration could be employed in the framework of an alternative FFD policy. However, the proposed scheme manages to capitalize on the benefits of the FFD algorithm, while forgoing any data movement.

In the VM scheduling scheme described above energy-efficient allocation of VMs can be achieved, without necessitating the migration of loads. However, the sequential allocation of VMs could result (during heavy load) in the allocation of VMs in remote DC servers, reserving them for a long period of time while all adjacent servers have returned to hibernation. Therefore, a safeguard is required restoring the system to its original state (i.e., a consolidated, sequential VM allocation) after a certain period of time. In this course, the system safeguard can adjust its restoration frequency dynamically, based on its previous performance. To elaborate, a list of previous restoration frequencies can be maintained along with the relevant energy savings, incurred by the system restoration. Each restoration frequency can then be assigned a weight based on the achieved energy savings and the current restoration frequency can be computed as the weighted average of all restoration frequencies in the list.

The proposed FFD scheme employing the above safeguard could be implemented through SDI, yielding significant energy savings in the framework of a DC interconnected ecosystem, without the employment of any load optimization algorithm. Most importantly such a SDN approach could reduce energy consumption even further if employed complementary to a load optimization module.

12.5 Evaluation

The evaluation procedure can be summarized as a set of semi-random DC configurations containing a (each time) variable numbers of DC rooms, racks, servers, and VMs, the number of the latter changing as a function of average server CPU and/or RAM utilization. In this sense and for the rest of this section, the term "DC configuration" will refer to a static number of:

1. DC rooms;
2. Racks per DC room;
3. Servers per rack;

4. CPU/RAM utilization.[1]

For each configuration, 20 emulations were performed in order to reduce the effect of randomness in the generation of VMs and servers' characteristics and the respective measurements. In each emulation, the characteristics (number of logical CPU cores, memory capacity, HDD capacity, consumption characteristics and whether it is a green one or not) of the servers change randomly bound by preconfigured minimum and maximum values, altering the DC configuration as to its computing power and energy consumption characteristics. Similarly, each time, four different VM flavors (e.g., VM virtual hardware configurations) are defined in a random, partially preconfigured, manner and the instantiated VMs follow the specifications of one of the generated flavors, at random. Based on the VMs characteristics, a set of semi-random[2] measurements were determined and corresponded to the VMs of each DC configuration.

12.5.1 Models Used

To calculate the energy savings incurred by the application of DOLFIN in the various DC configurations in a homogeneous manner, specific energy models were considered as to the energy consumptions of the physical DC servers, under the assumption that the servers were 100% dedicated to serving virtualization services, namely running applications residing inside VMs. Assuming a server s and a VM residing in this server, the following equations overview the energy models used:

$$E_s = a + b \cdot load_{s,cpu} + c \cdot load_{s,mem} + d \cdot load_{s,net} \tag{12.1}$$

$$E_v = b \cdot load_{v,cpu} + c \cdot load_{v,mem} + d \cdot load_{v,net} \tag{12.2}$$

where

- a is the energy consumption of the server when idle;
- b is a coefficient correlating the average CPU load of the server to the respective energy consumed;
- c is a coefficient correlating the average memory load of the server to the respective energy consumed; and

[1] For the present evaluation, only average CPU utilization has been considered, though setting RAM utilization is also allowable through the evaluation framework settings.

[2] The semi-randomness is based on the following: for each VM, a pseudo-random numerical ID gets generated and is fed to a sine function to affect the respective period. Next, based on the current emulation time, a value between 0 and 1 is calculated and is multiplied by the CPU/RAM characteristics of the VM, as dictated by its flavor to get the semi-random, to get the CPU/RAM measurements.

- d is a coefficient correlating the average network bandwidth utilized by the server to the respective energy consumed.

and

- $load_{s,cpu} = \sum_{v\in} load_{v,cpu}$
- $load_{s,mem} = \sum_{v\in} load_{v,mem}$
- $load_{s,net} = \sum_{v\in} load_{v,net}$

In addition to the energy models, in order to be able to evaluate the revenue-related performance of DOLFIN simultaneously allowing for the definition and support of flexible SLAs based on performance (in terms of "greenness" and computational capabilities provided to each user), a simple revenue model has been adopted, applied in the course of the optimization procedure. The revenue model includes the calculation of the possible earnings of the DC operators due to service (computational resources) provisioning in the form of VMs and the cost that occurs due to the energy consumption of the various DC elements. In this framework, the total revenue model that has been adopted for the evaluation of DOLFIN, assuming that the DC features S servers hosting V VMs in total and supported by a number of N non-IT infrastructure elements (e.g., lighting, HVAC), is summarized by

$$Revenue = \sum_{v=1}^{V} (v.cpus \cdot v.server_{cpu\_frequency} * cpu_{mult} + v.ram \cdot mem_{mult}) \cdot price_{offset} - \sum_{s=1}^{S} energy\_consumption\,(s) - \sum_{n=1}^{N} energy\_consumption\,(n) \tag{12.3}$$

where

- cpu_{mult} and mem_{mult} are multipliers characterizing the contribution of the CPU and RAM usage to the price determination and
- $price_{offset}$ is a variable used for scaling the price to the current operational environment of the DC and also allowing for special pricing for individuals or special groups of users.

For the energy consumption of the servers, namely the second summation apparent in (12.3), the aforementioned energy models have been employed. In order to calculate the energy consumption of the non-IT elements (lighting and HVAC), we have used generic rules and assumptions;

- It has been assumed that each rack is lit by a single lighting element of average power dissipation equal to 50 W;
- It has been assumed that the energy needed to cool a server equals the BTUs of heat output it presents, using a variable server efficiency parameter ranging from 0.6 up to 0.9 [36]. We have also assumed a standard temperature difference between the outside world and the DC-internal one, so that any changes in the cooling energy consumption occurs as a function of the heat load produced by the physical servers.

Evidently, a different revenue model will result in different figures as to the performance of DOLFIN in terms of creating actual revenue out of energy efficiency or performance policy actuation (see next paragraph). However, this model has been chosen as an indicative case that enables smart and flexible SLA provisioning, based on the actual computational and energy efficiency characteristics that are being provided by the DC operators to their clients.

12.5.2 Evaluation Outcome

In this section, the core conclusions from the DOLFIN evaluations as a whole are drawn. For each DC type of interest to DOLFIN (spanning from large commercial DCs to smaller urban DCs and micro-DCs), the DOLFIN solution was evaluated against several DC configurations, the total number of evaluations been calculated by the following equation:

$$\begin{aligned} TOTAL = {} & MAX\_NOMIN\_NO \times (RACKS\_PER\_ROOM\_NO - RACKS\_PER\_ROOM\_NO) \\ & \times (SERVERS\_RACK - SERVERS\_RACK) \\ & \times ((MAX\_UTILIZATION\_SERVERS\_RAM - MIN\_UTILIZATION\_SERVERS\_RAM)/10) \\ & \times OPTIMIZER\_PER\_SETUP \times POLICIES\_NO\_SIMULATIONS \end{aligned} \tag{12.4}$$

A total of a number of more than 15,000 simulations and (accordingly) optimization plans were conducted. The source code of the evaluation framework can be accessed online through the DOLFIN source code management platform [35], where details about the entire set of emulation parameters as well as instructions on how to configure the evaluation procedure are given. Indicatively, for the emulation processes in the context of urban micro-DCs operating 2–5 racks, the core configuration options are tabulated and presented in Table 12.1.

12.5.2.1 Urban Micro-DCs

In this case, we examine micro-DCs containing a very limited number of racks (2–5) in a single room. In the following, the performance of DOLFIN is examined on the basis of the above configuration (see Table 12.1), generating a number of 11,200 different scenarios; 20 repetitions per DC setup were conducted. Although we conducted the evaluation with the number of servers per rack changing from 2 up to 6, next the results for 5 servers per rack are presented, to facilitate the comparison with the forthcoming medium-sized DC results.

The following figures present the average number of VMs, the energy gain and the revenue benefit from applying DOLFIN when the initial micro-DC load changes from 20 up to 80% and the optimization policy has been set up (by the Policy Maker) to optimize against the energy consumption.

Table 12.1 Basic configuration of the core emulations set for the case of urban micro-DCs

Variable	Description	Value
MIN_ROOMS_NO	The minimum number of rooms per scenario	1
MAX_ROOMS_NO	The maximum number of rooms per scenario	2
RACKS_MIN_PER_ROOM_NO	The minimum number of racks per room	2
RACKS_MAX_PER_ROOM_NO	The maximum number of racks per room	5
SERVERS_MIN_PER_RACK	The minimum number of servers per rack	2
SERVERS_MAX_PER_RACK	The maximum number of servers per rack	6
SERVERS_MIN_RAM_GB	The minimum possible amount of RAM of a server in GB	16
SERVERS_MAX_RAM_GB	The maximum possible amount of RAM of a server in GB	128
SERVERS_MIN_CPU_CORES	The minimum possible number of logical cores of a server	16
SERVERS_MAX_CPU_CORES	The maximum possible number of logical cores of a server	24
SERVERS_MIN_FREQ	The minimum possible maximum frequency of a server in GHz	1.8
SERVERS_MAX_FREQ	The maximum possible maximum frequency of a server in GHz	4.0
MIN_INITIAL_UTILIZATION_ SERVERS_CPU	The minimum initial CPU utilization of the servers in total (DC utilization)	20%
MAX_INITIAL_UTILIZATION_ SERVERS_CPU	The maximum initial CPU utilization of the servers in total (DC utilization)	80%
OPTIMIZER_REPETITIONS_ PER_SETUP	The number of repetitions to optimize a certain setup (number of rooms, racks, and servers under fixed initial aggregate DC utilization)	10
SERVERS_PERCENTAGE_ GREEN	The percentage of green-powered servers	10%

The results indicate that as the number of hosted VMs increases with the average DC utilization (Fig. 12.3), the energy gain decreases as the DC configuration options get less and the possibility to result in inactive servers to put them in sleep state decreases (Fig. 12.4). Interestingly, a simple linear regression analysis indicates that the rate of energy benefit reduction as a function of the average DC utilization is on average about 4.1% for each 10% of increase in the average DC utilization with an R2 value of 0.96. In absolute numbers, the percentage of the expected energy gain for highly underutilized micro-DCs was ranged between 65 and 75% whereas the respective numbers for highly utilized micro-DCs were much lower ranging from approximately 42% down to 37%.

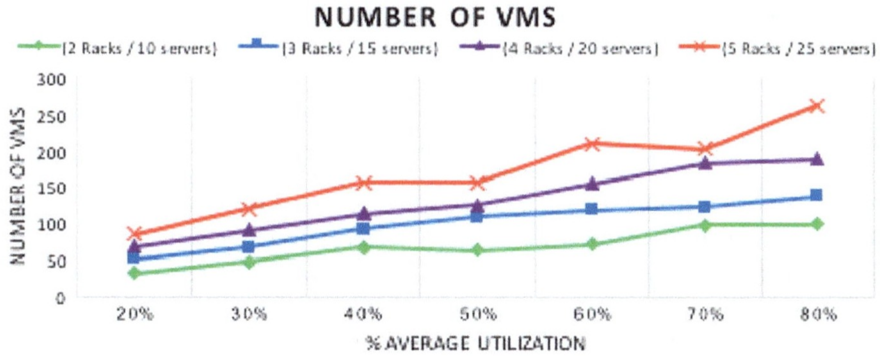

Fig. 12.3 Average number of VMs hosted by the Micro-DC as a function of the average micro-DC utilization

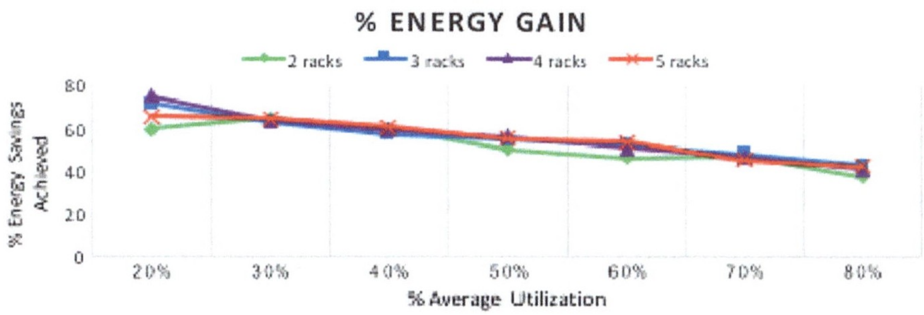

Fig. 12.4 Average energy consumption benefit as a function of the micro-DC utilization (energy efficiency policy)

Simultaneously with the decrease in the energy benefit as the DC utilization increases, the expected revenue change also decreases (Fig. 12.5), at a rate similar to the energy reduction one, presenting on average a benefit reduction rate of approximately 5.6% with an average R^2 value of approximately 0.9. In absolute numbers, the expected energy gain ranged from approximately 45–55% for the case of highly underutilized DCs down to 16–18% for the case of highly utilized DCs. This behavior is to be expected as the limited number of servers (thus limited heterogeneity on the hardware energy consumption) in combination with the policy applied the optimization procedure which was set to reduce the energy consumption of the micro-DC can explain the strong dependence between the two values.

Indeed, the following table presents the results of a simple correlation analysis between the expected energy and the expected revenue gain after the application of DOLFIN, the policy being set to optimize against energy consumption minimization. The high correlation value (0.91) indicates that the two attributes are highly correlated, the revenue benefit being caused by the lowered energy consumption of the DC elements (Table 12.2).

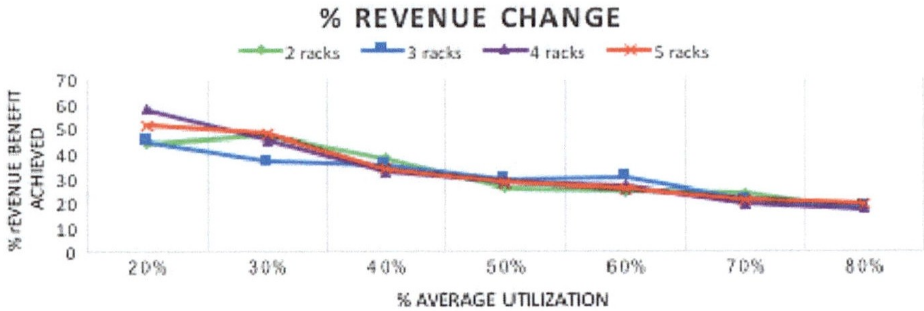

Fig. 12.5 Average revenue benefit as a function of the average micro-DC utilization (energy efficiency policy)

Table 12.2 Correlation analysis between the energy and the revenue gain for the case of 2 racks (energy policy)

	Energy gain	Revenue gain
Energy gain	1	0.91
Revenue gain	0.91	1

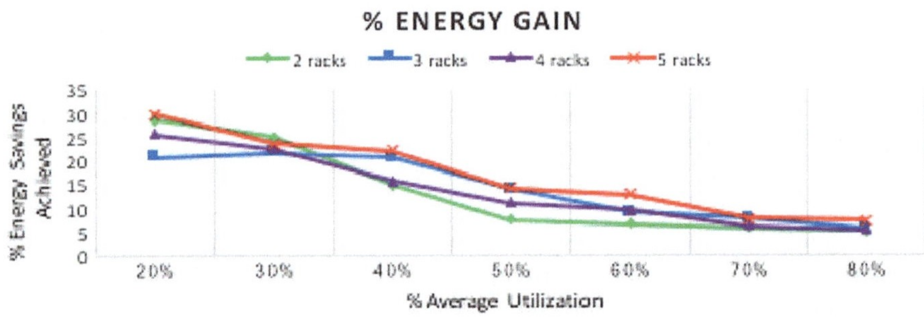

Fig. 12.6 Average energy consumption benefit as a function of the average micro-DC utilization (performance policy)

Next and as regards the DOLFIN system operation under a policy set to maximize the performance (hence revenue based on a revenue model that linearly correlates the revenue amount on the computational power offered to the DC clients). The following figures summarize the relevant findings (Figs. 12.6 and 12.7).

As expected, as DOLFIN did not account for prioritizing energy efficiency during this set of emulations, the percentage of the energy gain attained was significantly lower than when optimizing against energy efficiency. Moreover, the rate of change as a function of the average DC utilization dropped to an average of approximately 3.6% for each 10% increase in the average DC utilization, indicating that when operating under the "Performance" policy, the average utilization is of lesser importance; it is actually the heterogeneity of the energy characteristics of the more powerful (from a

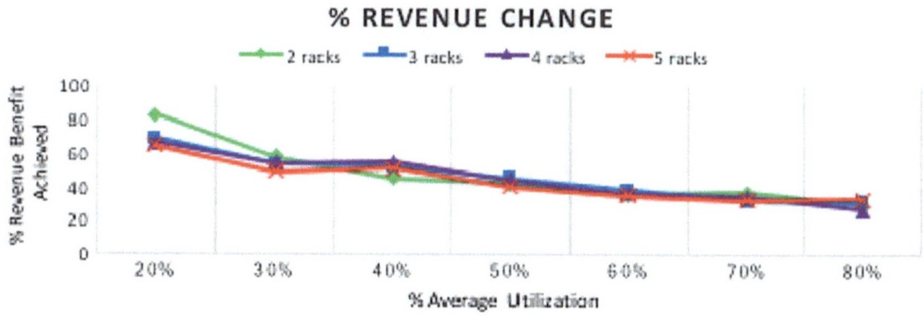

Fig. 12.7 Average revenue benefit as a function of the average micro-DC utilization (performance policy)

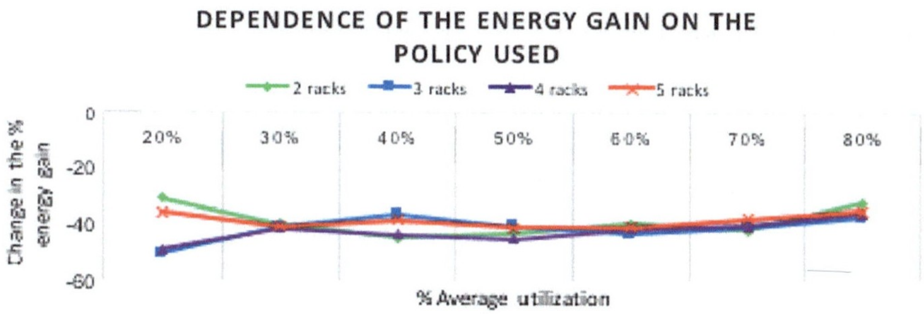

Fig. 12.8 Difference between the expected energy consumption change when optimizing against performance instead of energy efficiency

computational point of view) IT equipment that plays the most important role. Simultaneously, the absolute numbers of the expected energy benefit also dropped in by 40% (in absolute numbers) compared to the energy efficiency-oriented optimization case. The following figure graphically presents the above.

As apparent from Fig. 12.8, the energy consumption reduction when optimizing against energy efficiency targets significantly outpaces the respective when optimizing for DC performance; the average change on the anticipated energy benefit is approximately 40.1%.

Despite the significant drop in the expected energy benefits, the expected revenue benefit in the two cases is on the positive side. This is presented in Fig. 12.9, where the dependence of the revenue change on the policy used is depicted; despite the changes depending on the DC configuration and utilization, the average change on the anticipated revenue is, in total, 13.8%.

Finally, despite the change of in the absolute numbers of energy efficiency and revenue change when employing different policies, the correlation between the change in the energy gains and the expected revenue gains remains significant as deduced from Table 12.3.

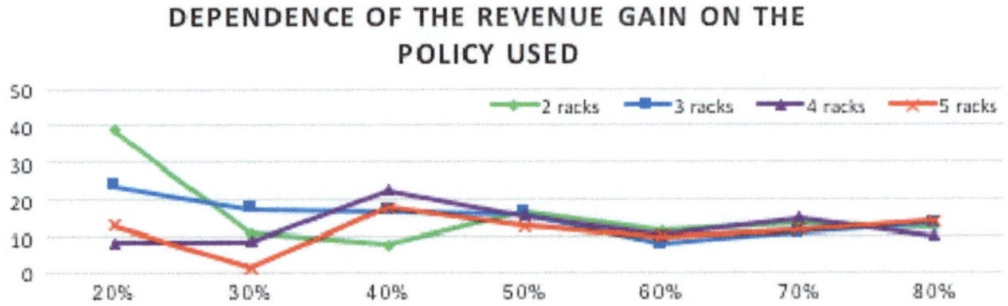

Fig. 12.9 Difference between the expected revenue change when optimizing against performance instead of energy efficiency

Table 12.3 Correlation analysis between the energy and the revenue gain for the case of 2 racks (performance policy)

	Energy gain	Revenue gain
Energy gain	1	0.91
Revenue gain	0.87	1

12.5.2.2 Urban Medium-Sized Urban DCs

In contrast to small-sized DCs that normally feature 2–10 racks with an average of 5 being the rule of thumb, medium-sized urban DCs are generally considered to contain 6–80 racks with an average number being 25. In the following, the performance of DOLFIN is examined on the same basis as the above configuration, though the number of racks was configured to be between 20 and 50, generating a number of 860 extra simulations; as in the case of the micro-DCs, 20 repetitions per DC setup were conducted. The number of servers per rack was fixed to five (5), hence the cases of 100, 150, 200, and 250 servers were considered. Last, for this set of emulations, we considered that approximately 10% of these servers (chosen at random) were powered by green sources, thus not contributing to the overall DC (brow) energy consumption.

Figure 12.10 presents the number of VMs that were considered, on average, for each DC setup:

Next, we set the DC policy to optimize against energy efficiency and calculated the average energy and revenue benefits that were expected to be acquired after the application of the generated optimization plans. The results from this set of emulations are presented in Figs. 12.11 and 12.12.

Comparing with the micro-DC case, it can be easily deduced that average expected energy benefit as seen in Fig. 12.11 is significantly larger than in the case of the micro-DCs when the DC is highly utilized; this can be attributed to two things:

1. The increased number of available configurations, as it allows for more proper optimization, exploiting at the highest possible extent the heterogeneity of the servers available;

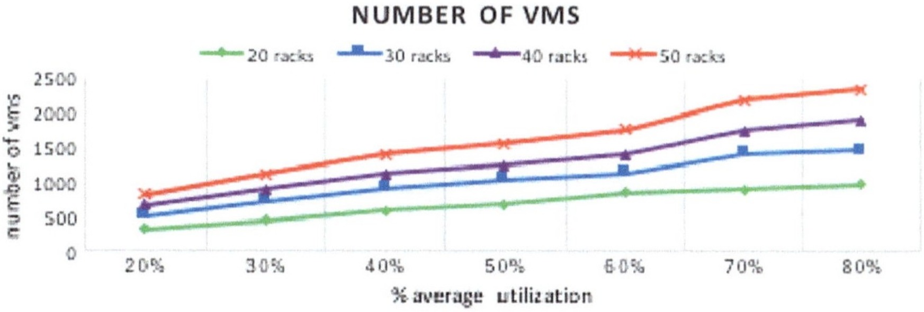

Fig. 12.10 Average number of VMs hosted by the Micro-DC as a function of the average medium—sized DC utilization

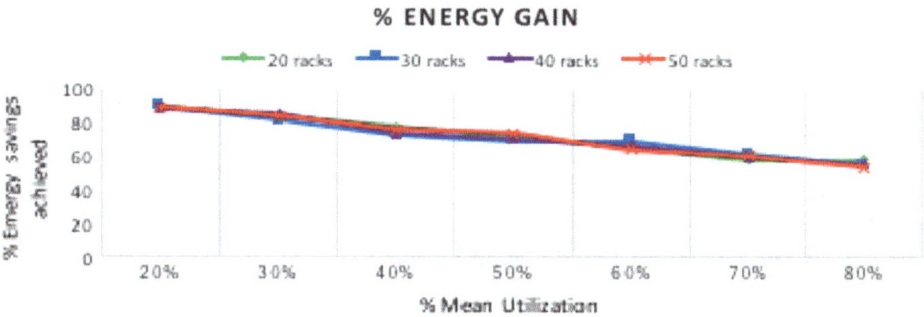

Fig. 12.11 Average energy consumption benefit as a function of the average medium-sized DC utilization (energy efficiency policy)

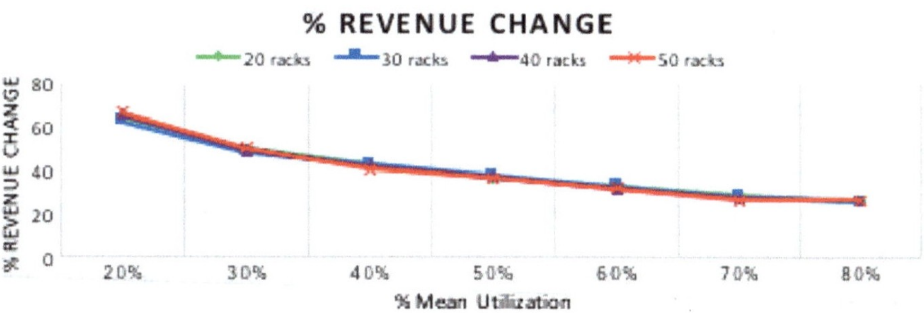

Fig. 12.12 Average revenue benefit as a function of the average medium-sized DC utilization (energy efficiency policy)

2. The introduction of the green-powered servers, as the optimizer attempts to load them as much as possible;

Interestingly and in accordance with the aforementioned two differentiating factors, the rate of energy benefit reduction (as calculated through a simple linear regression) is approximately 5.3% per 10% of increase in the mean DC utilization (with an R^2 value of 0.96), which is significantly lower than in the case of micro-DCs

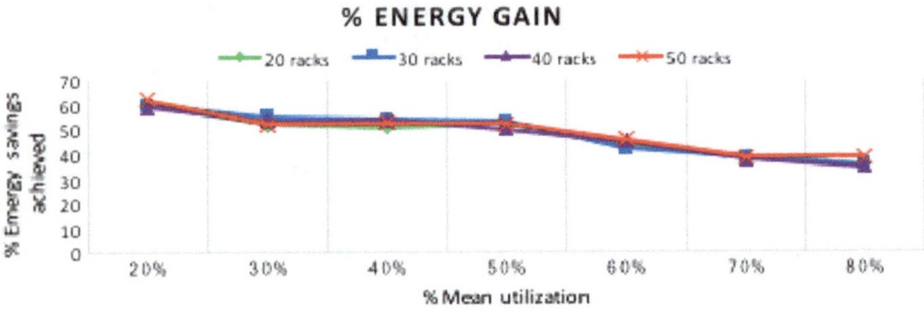

Fig. 12.13 Average energy consumption benefit as a function of the average medium-sized DC utilization (performance policy)

indicating that DOLFIN can yield very significant energy benefits in partially green-powered medium-sized DCs even if they are highly utilized.

Similar remarks hold for the average expected revenue benefit from the application of DOLFIN, where the expected revenue reduction per 10% of increase in the average DC utilization is approximately 5.77% with an R^2 value of 0.94 (on average).

As regards the correlation between the energy efficiency benefit and the revenue benefit, a very high correlation indicator was calculated reaching 0.97. Next, we set the DC policy to optimize performance and the respective results are drawn in the following figures.

As can be easily deduced by a simple examination of Figs. 12.13 and 12.14, once again, the actual scale of the DC in terms of number of racks does not significantly alter the performance of the DOLFIN optimization. The average rate of change of the energy benefit with respect to 10% of increase in the mean DC utilization is on average approximately 3.8% with an R^2 value of 0.90. Similar deductions can be made also for the examination of the expected revenue change as the average DC utilization increases, the respective rate of change being on average 5.24% with an R^2 value of 0.89.

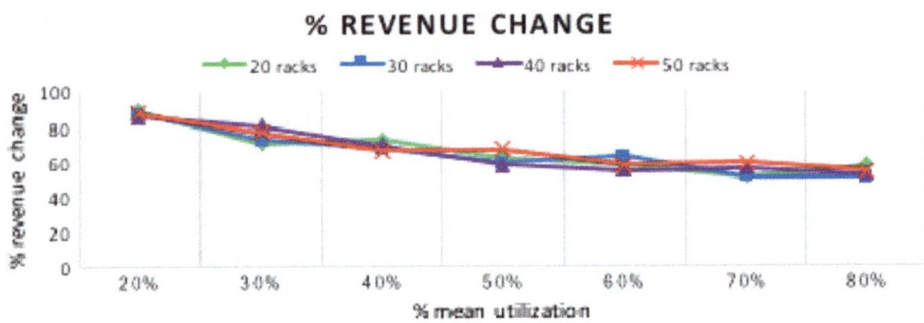

Fig. 12.14 Average revenue benefit as a function of the average medium-sized DC utilization (performance policy)

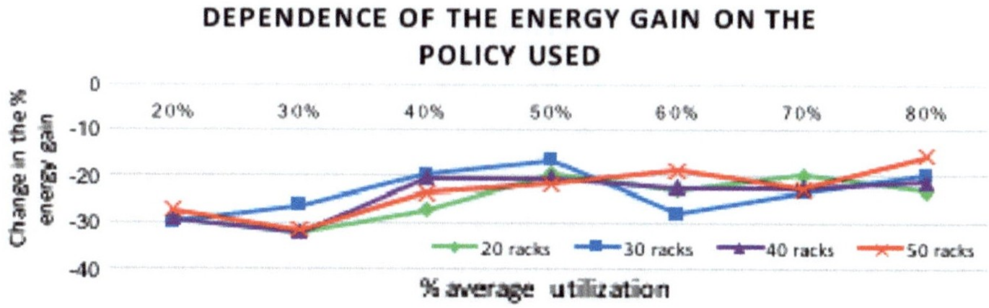

Fig. 12.15 Difference between the expected energy consumption change when optimizing against performance instead of energy efficiency (medium-sized DC)

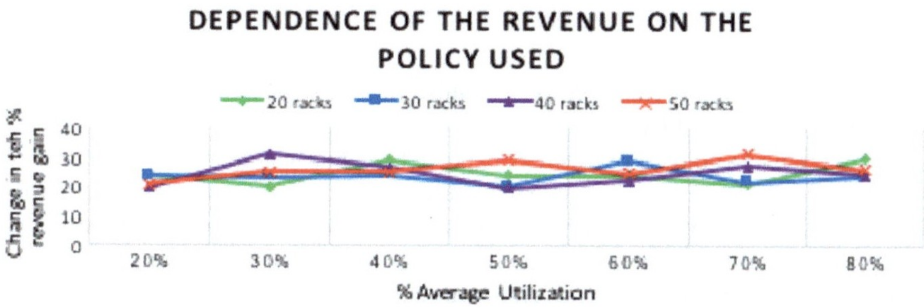

Fig. 12.16 Difference between the expected revenue change when optimizing against performance instead of energy efficiency (medium-sized DC)

Based on the above results, Fig. 12.15 depicts how the expected energy benefit changes as the policy changes (from energy efficiency-oriented to performance-oriented), whereas Fig. 12.16 presents the respective results focusing on the expected revenue change .

As expected, the expected energy benefit was significantly lower for the case of performance-based optimization (hence the negative values in Fig. 12.15) while, simultaneously, the expected revenue was significantly increased as a result of the higher value of the services offered. The average expected drop in the expected energy revenue reached −23.8% whereas the respective increase for the revenue metric was 24.68%.

Comparing the medium-sized with the micro-DC case, it is evident that the DOLFIN impact on the latter is much more significant than in the former. This can be attributed to the increased flexibility that the increased number of physical servers delivers. Of course, the presence of the green-powered servers also plays a very important role in the exhibited increase of the expected energy and revenue benefits identified in the context of the medium-sized urban DCs. However, it should be highlighted that despite the context, it is easily deduced that

1. The DOLFIN solution is able to scale to both types of urban DCs;
2. Its performance is almost linear to the level of average DC utilization;

3. It is able to be configured to favor energy efficiency over performance (hence revenue) and vice versa.

It should be also noted that;

1. A different pricing/revenue model would result in different evaluation outcomes in terms of revenue analysis;
2. In actual smart grid conditions where the price changes due to the smart grid Operator instructions might be higher, different results would also be attained.

12.6 Conclusions

A practical approach towards achieving energy-efficient data centers operation in the context of a relevant DC federation is presented. The energy profile optimization follows a spiral approach, pertaining four different optimization levels and scaling from single server to federation level. Several techniques for actualizing this energy consumption profile optimization of the federation of DCs are examined and discussed. The available degrees of freedom of the general problem of optimizing the energy profile of federated data center networks are classified and elaborated. Under this perspective, the developments of two relevant EU projects are examined, both considering DCs as active smart city actors, offering stabilization services and, even, energy to the smart grid supporting their operation. A simple technique for efficiently rearranging workload in the context of a single DC under heavy and light load is discussed and a number of forward-looking insights are provided. The proposed integrated DOLFIN approach has been evaluated in the framework of over 15,000 simulations, exhibiting extremely favorable results.

References

1. Data Centre Dynamics (Online). http://www.dataCentredynamics.com/research/market-growth-2011-2012
2. Index Mundi (2011) Historical data graphs per year (Online). http://www.indexmundi.com/g/g.aspx?v=81&c=us&l=en
3. Katz R (2009) Tech titans building boom: Google, Microsoft, and other internet giants race to build the mega data centres that will power cloud computing. IEEE Spectr
4. Hamilton J Internet-scale service efficiency (Online). http://research.microsoft.com/~JamesRH
5. Bash C, Forman G (2007) Cool job allocation: measuring the power savings of placing jobs at cooling-efficient locations in the data centre
6. Multi-Tenant Data Centers Need To Play Bigger Energy Efficiency Role (Online). http://goo.gl/vplg4v
7. Zhuravlev S, Saez J, Blagodurov S, Fedorova A, Prieto M (2013) Survey of energy-cognizant scheduling techniques. IEEE Trans Parallel Distrib Syst 24(7):1447–1464
8. Orgerie AC, De Assuncao MD, Lefevre L (2014) A survey on techniques for improving the energy efficiency of large scale distributed systems. ACM Comput Surv 46(4)

9. Weiser M, Welch B, Demers A, Shenker S (1994) Scheduling for reduced cpu energy. In: Proceedings of the 1st USENIX conference on operating systems design and implementation
10. Ge R, Feng X, Chun Feng W, Cameron K (2007) Cpu miser: a performance-directed, run-time system for power-aware clusters. In: International conference on parallel processing, ICPP 2007
11. Kim W, Gupta M, Wei G-Y, Brooks D (2008) System level analysis of fast, per-core DVFS using on-chip switching regulators. In: IEEE 14th international symposium on high performance computer architecture
12. Le Sueur E, Heiser G (2010) Dynamic voltage and frequency scaling:the laws of diminishing returns. In: Proceedings of the 2010 international conference on power aware computing and systems, Berkleley, CA, USA
13. Lee YC, Zomaya A (2011) Energy conscious scheduling for distributed computing systems under different operating conditions. IEEE Trans Parallel Distrib Syst 22(8):1374–1381
14. David H, Fallin C, Gorbatov E, Hanebutte UR, Mutlu O (2011) Memory power management via dynamic voltage/frequency scaling. In: Proceedings of the 8th ACM international conference on autonomic computing, New York, NY, USA
15. Deng Q, Meisner D, Bhattacharjee A, Wenisch TF, Bianchini R (2012) Coscale: coordinating cpu and memory system DVFS in server systems. In: 2012 45th annual IEEE/ACM international symposium on microarchitecture
16. Mahadevan P, Banerjee S, Sharma P, Shah A, Ranganathan P (2011) On energy efficiency for enterprise and data center networks. Commun Mag 49(8):94–100
17. Chabarek J, Sommers J, Barford P, Estan C, Tsiang D, Wright S (2008) Power awareness in network design and routing. In: The 27th conference on computer communications (INFOCOM)
18. Sohan R, Rice A, Moore A, Mansley K (2010) Characterizing 10 gbps network interface energy consumption. In: 2010 IEEE 35th conference on local computer networks (LCN)
19. Heller B, Seetharaman S, Mahadevan P, Yiakoumis Y, Sharma P, Banerjee S, McKeown N (2010) Elastictree: saving energy in data center networks. In: Proceedings of the 7th USENIX conference on networked systems design and implementation
20. Mahadevan P, Sharma P, Banerjee S, Ranganathan P (2009) Energy aware network operations. In: Proceedings of the 28th IEEE international conference on computer communications workshops, Piscataway, NJ, USA
21. Chen G, He W, Liu J, Nath S, Rigas L, Xiaom L, Zhao F (2008) Energy-aware server provisioning and load dispatching for connection-intensive internet services. In: Proceedings of the 5th USENIX symposium on networked systems design and implementation, Berkeley, CA, USA
22. Voorsluys W, Broberg J, Venugopal S, Buyya R (2009) Cost of virtual machine live migration in clouds: a performance evaluation. In: Proceedings of the 1st international conference on cloud computing
23. Beloglazov A, Buyya R (2010) Energy efficient allocation of virtual machines in cloud data centers. In: 2010 10th IEEE/ACM international conference on cluster, cloud and grid computing (CCGrid)
24. Ghribi C, Hadji M, Zeghlache D (2013) Energy efficient vm scheduling for cloud data centers: exact allocation and migration algorithms. In: 2013 13th IEEE/ACM international symposium on cluster, cloud and grid computing (CCGrid)
25. Mandal U, Habib MF, Zhang S, Mukherjee B, Tornatore M (2013) Greening the cloud using renewable-energy-aware service migration. IEEE Netw 27(6):36–43
26. DOLFIN Project Number: 609140; Strategic objective: FP7-SMARTCITIES-2013(ICT-2013.6.2) (Online). http://www.dolfin-fp7.eu/
27. GEYSER Project Number: 609211; Strategic objective: FP7-SMARTCITIES-2013(ICT-2013.6.2) (Online)
28. Bash C, Forman G (2007) Cool job allocation: measuring the power savings of placing jobs at cooling-efficient locations in the data centre. In: ATC'07 2007 USENIX annual technical conference on proceedings of the USENIX annual technical conference

29. Bohrer P, Elnozahy EN, Keller T, Kistler M, Lefurgy C, McDowell C, Rajamony R (2002) The case for power management in web servers. In: Power aware computing, Kluwer Academic Publishers, Norwell, MA
30. Fan X, Weber W-D, Barroso LA (2007) Power provisioning for a warehouse-sized computer. In: Proceedings of the ACM international symposium on computer architecture, San Diego, CA, USA
31. Garey MR, Johnson DS (1979) Computers and intractability: a guide to the theory of NP-completeness. W.H.Freeman and Company, New York, NY
32. Openstack Foundation Openstack (Online)
33. OpenNebula OpenNebula (Online)
34. Eucalyptus (Online)
35. DOLFIN Evaluation framework source code (Online). https://stash.i2cat.net/scm/dol/evaluation_framework.git
36. APC Calculating total cooling requirements for data centers (Online). http://apcmedia.com/salestools/nran-5te6he/nran-5te6he_r3_en.pdf. Accessed July 2016

Chapter 13
Improving the Energy Efficiency by Exceeding the Conservative Operating Limits

Lev Mukhanov, Konstantinos Tovletoglou, Georgios Karakonstantis, George Papadimitriou, Athanasios Chatzidimitriou, Manolis Kaliorakis, Dimitris Gizopoulos and Shidhartha Das

13.1 Introduction

As discussed in previous chapters, the explosive growth of Internet-connected devices and the resulted flood of generated data will soon increase the demand for more powerful computing resources in data centres in support of the Cloud and Edge computing paradigm [1, 2]. However, such an increase of resources can only be realized if the strict data center energy budgets can be satisfied. Staying under the set energy budget is not only important for allowing operation of all the resources and for environmental sustainability but also for maintaining the operational costs low. In fact, according to Amazon's estimates [3], at its data centers, expenses related to the cost and operation of the servers account for 53% of the total budget, while energy-related costs amount to 42% of the total. In addition, development of servers with improved energy efficiency is detrimental for the realization of emerging decentralized data centers at the Edges of the Internet, where cooling infrastructure and power supply options cannot be as complex as in traditional centralized (Cloud) data centers [4]. However, improving the energy efficiency of servers is extremely challenging due to the stagnant voltage scaling (the most effective power saving knob), and the worsening process variations that nanometer circuits are experiencing [5–7], thus there is a need for efficient ways to deal with those issues.

L. Mukhanov · K. Tovletoglou · G. Karakonstantis (✉)
ECIT Institute, Queen's University Belfast, CSB Building, 18 Malone Road, Belfast BT9 6RT, UK
e-mail: g.karakonstantis@qub.ac.uk

G. Papadimitriou · A. Chatzidimitriou · M. Kaliorakis · D. Gizopoulos
University of Athens, Athens, Greece

S. Das
ARM Ltd., Cambridge, UK

C. Kachris et al. (eds.), *Hardware Accelerators in Data Centers*,
https://doi.org/10.1007/978-3-319-92792-3_13

13.2 Challenges in Nanometer Era

As the integrated circuits technology moves into the nanometer regime, circuits have started experiencing dramatically deteriorating effects of material properties on (active and leakage) power and die yields. In particular, the inaccurate manufacturing processes led to substantial fluctuations in critical device/circuit parameters of the manufactured parts across the die, die-to-die and over time which are only worsening as critical transistor dimensions reach the atomic scale [8]. The consequence is that the microelectronic substrate on which modern computers are built is increasingly plagued by variability in performance (speed, power) and fault characteristics, both static across multiple instances of a system and dynamic over its usage life. An immediate impact of such variability is on-chip yields; a growing number of parts are thrown away since they do not meet the timing and power-related specifications. The dynamic variations such as voltage droops and aging are extremely threatening for the correct operation of all the essential components of today's computing systems, i.e., processors, static memories used as caches and dynamic memories [8].

Another factor that worsens the power and performance variability is the packing of millions of transistors in a single die in line with the design trend of multi-core processors followed for continuing the performance scaling in support of the multifunctional capabilities of the popular consumer gadgets. Most importantly, such high transistor congestion is increasing the power consumption that is becoming a predominant concern especially after the end of Dennard scaling in mid-2000, which guaranteed that the power density will stay constant as transistors were getting smaller. The stagnant power scaling has led to the threat of dark silicon according to which the performance scaling in future multi-core systems is power limited in such a degree so that at the 7 nm process node (2018), more than 50% of the transistors in a processor will have to be powered off in every cycle [9].

13.3 Conventional Low-Power Variation-Aware Design

Various techniques have tried to address the above challenges in isolation. On one side, techniques for dealing with transistor variability have involved extra provisioning in circuits, known as *guard bands*, to account for the expected performance degradation of transistors and potential functionality failures. While such guard bands (in terms of increased voltage margins, circuit redundancy) have successfully ensured reliable operation up to date, their effectiveness in detecting and correcting all possible errors is being doubted by researchers, as geometries and supply voltages are being scaled down and circuits become more vulnerable to failures. As an example, the voltage guard bands (i.e., voltage up-scaling) currently adopted against a variety of issues are already significant ranging from at least 5% for die-to-die variations to 20% for addressing voltage droops. Indicatively, recent measurements in ARM processors indicated more than 30% timing and voltage margins in 28 nm [10].

Recent studies have also revealed that the refresh-rate adopted in dynamic memories (DRAMs) is extremely pessimistic [11, 12] and can be relaxed beyond the conservative 64ms that is currently adopted in DDR3 technologies. Such voltage, frequency, and refresh-rate margins are becoming more prominent with the use of more cores per chip, the increased voltage droops, reliability issues at low voltages (Vmin), and core-to-core variations [13–15]. All in all, it is becoming apparent that the amount of redundancy and guard banding that may be necessary to protect all circuits against the potential massive errors will soon lead to paramount energy overheads not sustainable by future systems, thus conflicting with the other major challenge, energy dissipation.

Power management techniques such as supply voltage scaling, power gating, multiple-supply (V_{DD}), and threshold (V_{TH}) voltage schemes have tried to limit the energy dissipation up to now. Among them, the by far most effective energy reduction technique is the reduction of the supply voltage while compensating for the performance loss with additional hardware resources [8]. In extreme cases, this strategy has been applied up to the point where circuits operate below the threshold voltage of their transistors (i.e., in their "almost-off" state). Unfortunately, voltage scaling may reduce power consumption significantly but increases the mean of the path delay distribution, as well as its variance, thus worsening the effect of parametric variations and making the circuits even more prone to erroneous and unpredictable behavior. Hence, in turn, even more protection and pessimistic assumptions are required to counteract the effects leading to a vicious circle between energy efficiency and reliability that ultimately diminishes the returns from technology scaling [16].

13.4 Beyond Pessimistic Variation-Aware Design

Triggered by the deadlock between reliability and energy reduction techniques, researchers have tried to overcome the conventional pessimistic measures applying techniques at the circuit/architecture that could allow aggressive voltage scaling while tolerating errors induced by violating safety margins [8, 17]. One of the first attempts was *Razor*, a processor design, which is based on dynamic detection and correction of timing failures of the critical paths due to below-nominal supply voltage [18]. The key idea is to tune supply voltage by monitoring the error rate during operation using shadow latches controlled by delayed clocks. By comparing the values latched by the main flip-flop and the shadow latch, an error condition could be detected, and the value of the shadow latch which is guaranteed to be correct is then used to correct the failure. Average energy gains of up to 38% for a series of SPEC2000 benchmarks were reported. However, main limitation of such a technique is the performance penalty that might incur in case of frequent failures, especially under low voltages, hindering their applicability to parallel data paths and demanding signal processing applications.

An experimental processor from Intel places tunable replica circuits and employs error-detection sequential [19] within the critical paths of the pipeline stages for detecting dynamic variations. Once a timing error is detected, the core prevents the errant instruction from corrupting the architectural state and initially flushes the pipeline to resolve any complex bypass register issues. To ensure error recovery, the core supports instruction *replay* at half frequency and multiple-issue instruction replay at the same frequency. A similar approach was also applied in [20].

Researchers have only recently started to venture in new error-resilient paradigms that relax the strict enforcement of precise hardware functionality, trading off reliability and quality of service/results (QoS) for lowering voltage by exploiting the inherent error resiliency of many algorithms [8, 21]. Main drive for such a paradigm shift is the fact that many algorithms are often statistical and iterative implying that errors can easily get averaged out and successive iterations may even correct previously introduced errors imparting them with a self-healing nature. Works at different layers including the circuits, processors, memories, and programming frameworks [22–28] have tried to take advantage of the inherent error resiliency but such approaches are still targeting specific applications and cannot be generalized especially in server environments where workloads vary significantly.

Researchers have also tried to improve the fault tolerance of system software. Gu et al. [21] use fault injection to characterize the behavior of the Linux kernel in the presence of faults. The authors in [29] investigate mechanisms to heal the OS in the presence of faults, without rebooting and destroying the state of—potentially unharmed—applications. In the context of Hypervisors, FT-Xen [30] routes all writes to mutable state through a single core which is considered reliable. However, this requires extensive Hypervisor modifications, unless the Hypervisor is inherently nonsymmetric.

Several recent efforts have tried to improve the fault tolerance of a data center by developing techniques to detect and predict the failures that may occur in a cloud datacenter. These techniques [31–34] generally leverage machine learning or statistical analysis techniques to process the log data generated from the physical or virtual servers to understand the causes of the past failures and use this information to detect and predict future failures in real-time. A failure prediction method is proposed in [25], for instance, for cloud data centers that use the pattern of the system log messages to predict a failure by classifying the messages by their similarities in real-time. All these, and similar other, techniques are independent of the cloud middleware and are not integrated with the latest available popular resource management frameworks such as OpenStack.

13.5 UniServer Approach—Operation at Extended Margins

As discussed above, recent works have tried to overcome the power and performance overheads imposed by the traditional pessimistic design paradigm but very few of them have been applied in the context of a server. In this section, a new paradigm called UniServer [42, 43] is being described.

Figure 13.1, depicts the different layers of the UniServer ecosystem. The most fundamental idea of the project lies on the hypothesis that each hardware component (i.e., core, cache, and DRAM) may have intrinsically different capabilities in terms of energy, performance, and reliability. Starting from the low layers, we develop techniques that aim at revealing new Extended Operating Points (EOP) for each hardware component based on the component's true capabilities. This is achieved by stress testing the hardware components during a pre-deployment phase under different points using various stress kernels. During deployment, a HealthLog daemon is monitoring the health status of the hardware under any used voltage/frequency/refresh rate (V-F-R) point and informs the system software by propagating information vectors about the performance, power, temperature, and any incurred errors. Moreover, another Linux daemon, the StressLog, is responsible for periodic offline, on-demand stress testing of the hardware components and for producing an output vector containing the new safe system V-F-R margins that will be suggested to the software (i.e., Hypervisor) for future use. It also produces log files recording errors (correctable

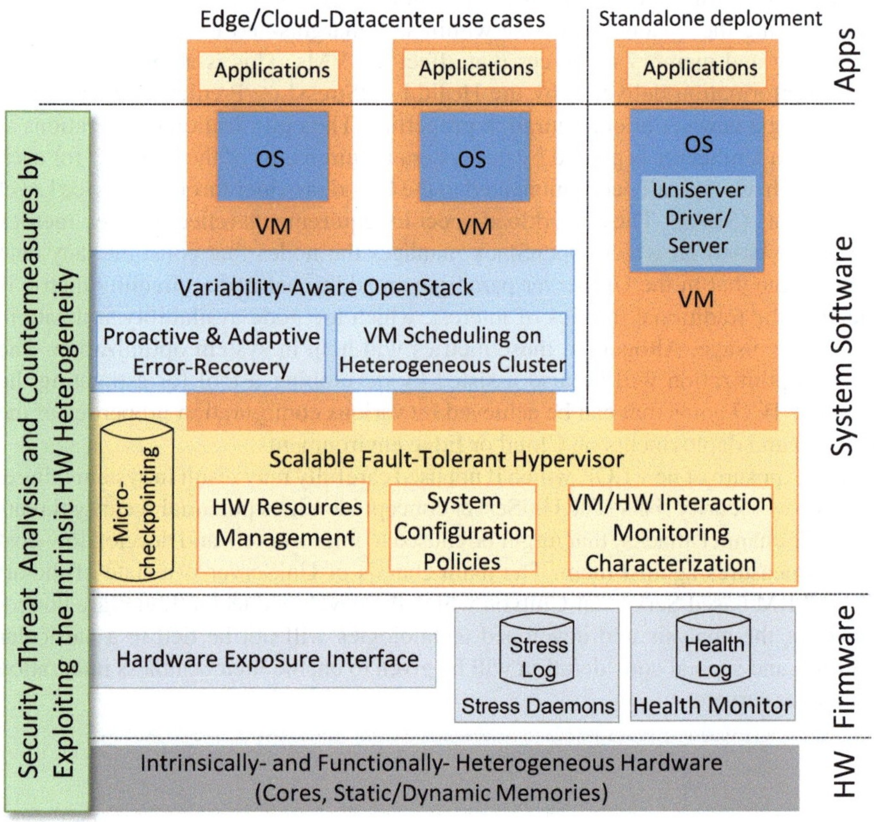

Fig. 13.1 Cross-layer error-resilient server ecosystem

or uncorrectable), system configuration values, sensor readings, and performance counters. Using the information provided by the HealthLog and StressLog the Predictor develops probability failure models and tries to predict the hardware behavior under any operating point and eventually helping the system software to decide on the optimum configuration.

The UniServer paradigm addresses a wide range of use cases, ranging from deployments in remote locations close to the end users to deployments in cloud data centers. To facilitate such diverse use cases, the UniServer platform must be equipped with a complete software stack that can efficiently manage any compute and storage resources by offering easy installation, migration, and replication of tasks, either at the node or server-rack level. To this end, state-of-the-art software packages for virtualization (Hypervisor) and resource management (OpenStack) are being adopted. Such packages, apart from managing the virtual machines (VMs) at the node level (Hypervisor) and the resources at a rack/data center level (OpenStack), they are also being enhanced for optimizing the system operation and the available resources by fine-tuning the extended V-F-R points. In particular, the Hypervisor will aim at limiting the effects of the potential faults to higher software layers by reconfiguring the system to operate within safe margins and isolating problematic processing and memory resources that affect the VMs. This is achieved by utilizing the information delivered by the HealthLog/StressLog/Predictor daemons and developing a new set of configuration properties. The optimization of operations at the EOP in UniServer is guided by the system requirements of the end-user for each VM, which are typically communicated to the Cloud provider through Service Level Agreements (SLAs). These workload-specific requirements reflect the key metrics of interest based on which OpenStack manages the nodes that constitute any data center. Note that in the UniServer paradigm an additional node reliability metric is added to the traditional metrics of interest, which are node availability, utilization, and energy usage. Altogether, these metrics will help in system optimization. The system optimization will be also assisted by developing a tool for estimating the potential TCO gains that can be achieved by various configuration properties of the platform and deployments on Cloud or Edge environments.

The exposure of new EOP, which if not used carefully may result in system failure, entails new security risks. The UniServer concept identifies potential security threats (i.e., side channel attacks) that might be caused to microservers and develop low-cost countermeasures against them. The main chassis of UniServer is a state-of-the-art 64-bit ARM based Server-on-Chip on which the developed technologies are ported. However, the analysis and developed technologies will not be tied to a particular platform and special consideration will be given to enable their seamless integration with other servers.

13.5.1 Exposing Margins and Monitoring Hardware Behavior

To reveal optimistic margins, there is series of tests that are planned before and after the server deployment. First, at the pre-deployment stage, the system goes through a batch of stress tests to determine the more efficient but safe per-component margins. Second, at normal operation in the field, a daemon is constantly recording any possible errors (even if correctable) to fine-tune the margins after deployment. If the number of errors rises above a certain threshold a new stress-test cycle may be triggered to determine new efficient safe margins. This is useful to better adapt to the workloads and also to the aging of the system. Third, during runtime, a predictor daemon is running to observe different metrics and advise the Hypervisor on possible execution modes (e.g., high-performance or low-power).

13.5.1.1 Revealing the Margins Within On-Board Components

Heterogeneity exists among cores located on the same chip, DRAM and cache memory banks. Each resource may perform better or worse than others but certainly not as any other similar resource on the board. In the UniServer paradigm, each core and memory bank is considered and characterized individually. For example, for each cache memory bank the minimum voltage that allows correct operation can be revealed. This information could be exploited by software to achieve better energy efficiency without compromising reliability.

13.5.1.2 Stress-Test Development

First of all, the underlying cores and memories will be stressed using diagnostic viruses. Genetic algorithms for generating these viruses are being used. These viruses cause maximum voltage noise, power consumption, and error rates. The viruses will represent a pathogenic worst-case scenario that is unlikely to be encountered in real-life workloads. Safety margins are more pessimistic than these worst-case viruses, therefore these stress tests will reveal initial EOP. In addition, real-life workloads will probably allow even more efficient margins since they produce significantly less voltage noise, power consumption, and error rates compared to stress viruses.

13.5.1.3 HealthLog Daemon

Operating outside the nominal values may introduce hardware errors during the system's lifetime. Thus, there is a need for a runtime mechanism that will monitor the system and report errors occurring during up-time. Such mechanisms already exist for different platforms [ref MCELOG] [ref LINARO] but their functionality

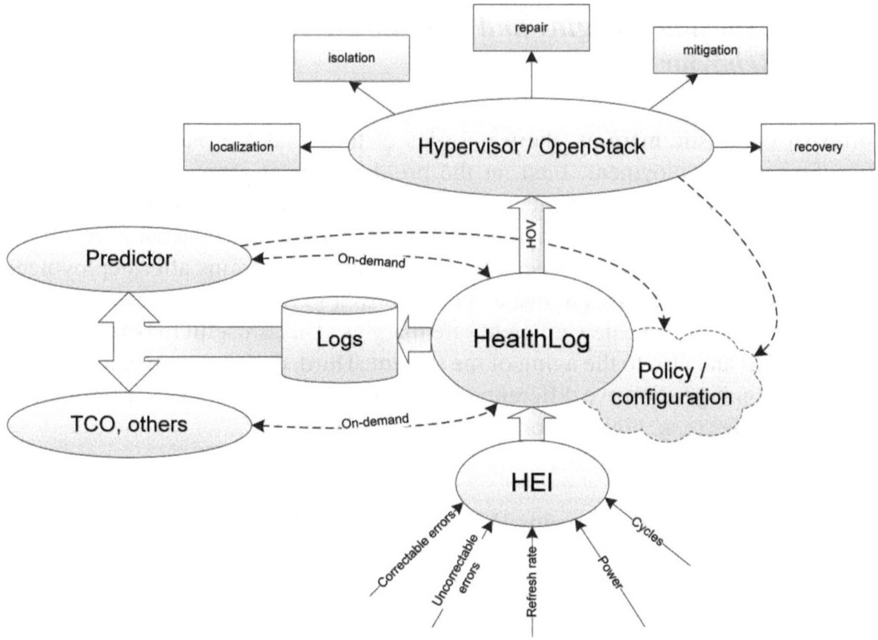

Fig. 13.2 HealthLog monitor

is limited on logging of the hardware events. The UniServer paradigm implies a specific monitoring mechanism that will react to hardware events and offer additional information of the system state. The mechanism which uses extended error and system monitoring capabilities, including various sensors and performance counters, is called the HealthLog monitor [47].

The HealthLog monitor will interact and exchange information (Fig. 13.2) with higher system layers (e.g., the Predictor and the Hypervisor) and it will also offer several logging options, such as different log levels and multiple logging formats, varying from raw text to XML and database compatible output. The HealthLog monitor will provide two types of services: (a) Event-driven services, where it will collect information based on event occurrences in the system (e.g., errors) and (b) On-demand services, where the monitor will respond to requests from higher layers for specific information.

The operation of HealthLog is required to initiate prevention and recovery procedures, and thus it must not introduce unnecessary delays and should be lightweight, without causing any additional system overhead. For this reason, HealthLog monitor offers configuration of event priorities and the optional capability of operating in kernel space, to further reduce the reaction time on critical errors. The extended features of HealthLog are still operated in user space and the module operation is limited to decode hardware errors and notify the hypervisor, in the case of critical errors.

In user space, HealthLog daemon can record error statistics, measure event rates and count their occurrences. Additionally, it offers the capability of extending list of events, introducing conditional events (or condition-met events). These events are a combination of counters and a described condition. For instance, a new conditional event can be “if L2-correctable errors are more than 10”. When the condition is met, depending on the current configuration, it will be logged and a process interested for this event can be notified.

HealthLog reacts to the event occurrences depending on its current configuration, which is described in the HealthLog policy. Every process can request logging or notification of events by supplying its own policy file on HealthLog daemon. The daemon will log the occurrence of events if at least one of the applied policies (including the default) has requested logging of that particular event. Similarly, HealthLog will notify the processes that requested a notification for an event.

The extended features and flexibility of HealthLog monitor can be used for both hardware characterization and fault resilient software operation. The monitor can be programmed accordingly to identify possible symptoms, which may not be necessarily errors and react by notifying the hypervisor about a possible upcoming failure.

13.5.1.4 StressLog Daemon

The UniServer paradigm implies changing the nominal V-F-R values to reduce the power consumption of each server in the system. These new values may need to be updated several times over the lifetime of a server due to the aging effects of the machine or unexpected errors observed. Therefore, a mechanism is needed to produce new nominal values that will still guarantee the safe operations of the server. This mechanism will stress the machine using predefined applications and compute new safe operating V-F-R margins. We call this mechanism the StressLog monitor. The StressLog monitor will be spawned either periodically during a machines lifetime (e.g., every 2–3 months) or will be triggered by higher system layers in the case of erratic or anomalous machine behavior. The machine being tested will be taken offline and as soon as the monitor receives the input stress target parameters from the higher system layers, it will initiate the stress test scenarios. The StressLog monitor will also include a workload suite, consisting of different benchmarks and kernels that either represent real-life applications or are hand-coded to stress-specific components of the system. During a stress test, the HealthLog monitor will run in parallel to record system events (errors, system values, sensors, and performance counters). The StressLog monitor will take the output of the HealthLog and will wrap the needed information (defined in the stress target parameters) into a vector to be passed to the higher system layers.

13.5.1.5 Predictor

In order to advise the system regarding the best V-F-R mode depending on the current workload and runtime characteristics of the system, we will develop a machine learning predictor that interacts with the HealthLog and StressLog monitors to provide advice to the Hypervisor for choosing the desired operation mode.

13.5.2 Managing Operation at Extended Margins at System Software

13.5.2.1 Virtualization

One of the major breakthroughs in the UniServer ecosystem is the ability to explore and allow operation when possible at EOP. In fact, such points may dynamically change depending on the workload, variations of environmental conditions, chip aging, etc., and thus the system should be able to decide on the best energy-efficient configuration parameters in a fast and reliable way. At the same time, operating so close to the points of failure requires mechanisms to deal with potential, inadvertently introduced faults. UniServer follows a Hypervisor-based approach based on KVM to leverage all benefits of virtualization, such as easier deployment, administration, replication, and migration which are necessary for the targeted data centers at the Edge of the Cloud.

In the UniServer paradigm, the Hypervisor has additional roles. It is responsible for creating an appropriate execution environment for VMs by manipulating the power/performance/reliability trade-offs in an educated and safe manner. Specifically, it sets the system at a just-right configuration, which reduces the power footprint of each node by eliminating unnecessary hardware guard-bands, without introducing negative effects on the services running within the VMs. The best configuration depends on a number of different parameters, including the characteristics of application software, the capabilities of the specific hardware parts at the specific time and under the specific environmental conditions, as well as the QoS requirements introduced by the cloud management framework (OpenStack).

Despite applying sophisticated configuration policies within the limits specified by the StressLog, sporadic errors may still inadvertently occur due to the elimination of guard bands. The Hypervisor needs to offer VMs a reliable virtual execution environment on top of potentially unreliable hardware. In other words, it needs to transparently mask errors from upper software layers. At the same time, it needs to protect the whole system from catastrophic failures. Being the lowest level of system software, the Hypervisor itself needs to be resilient to errors. Beyond selecting a realistic hardware configuration, the Hypervisor isolates problematic processing and memory resources experiencing high error rates, as reported by the HealthLog. This is exactly one of the main aims at the Hypervisor layer and probably less complex than the upper software layers. In particular, the Hypervisor will be enhanced with mechanisms to transparently mask errors from upper software layers, and protect the whole system from catastrophic failures while choosing the right EOP for any given condition/user requirement.

13.5.2.2 Resource Management—OpenStack

The next layer of software is the cloud computing platform. OpenStack [30] makes an ideal candidate for this layer as it is a widely used open source middleware for cloud setups, and it pairs well with the popular enterprise and open source technologies. Our extended version of OpenStack includes support for monitoring VMs and determining their dynamically changing characteristics and virtual resource utilization at a finer granularity than the existing state-of-the-art. This resource monitoring information will be exploited to design and develop new scheduling policies, as well as to assess the susceptibility of VMs to experience catastrophic errors due to hardware faults. The new scheduling policies will also focus on incurring minimal overhead and being non-intrusive in real-world scenarios where OpenStack would manage streams of incoming and terminating VMs. Developing such an error-resilient software stack will not only help to avoid system crashes even at EOP but will also help in characterizing and exploring the server operation at aggressive V-F-R scaling points by exploiting the characteristics of real world workloads. Furthermore, by porting the OpenStack on a micro-server will enable resource management capabilities from classical Cloud data centers at the Edge.

13.6 Potential Impact—Design Time Characterization

In this section, the potential impact in relaxing the voltage and refresh rate in CPUs and DRAMs as well as the efforts in estimating the impact of IR drop is presented.

13.6.1 Characterization of CPUs

The initial phase of the UniServer project is focused on revealing the margins in voltage an frequency in CPUs. To this end, an experimental evaluation was performed on the X-Gene 2 platform using an automated framework to accelerate the characterization process and provide reliable and detailed results [44–46]. The primary goals of the automated framework are (1) to identify the target system's limits when it operates at scaled voltage and frequency conditions, and (2) to record/log the effects of a program's execution under these conditions. The framework provides the following features:

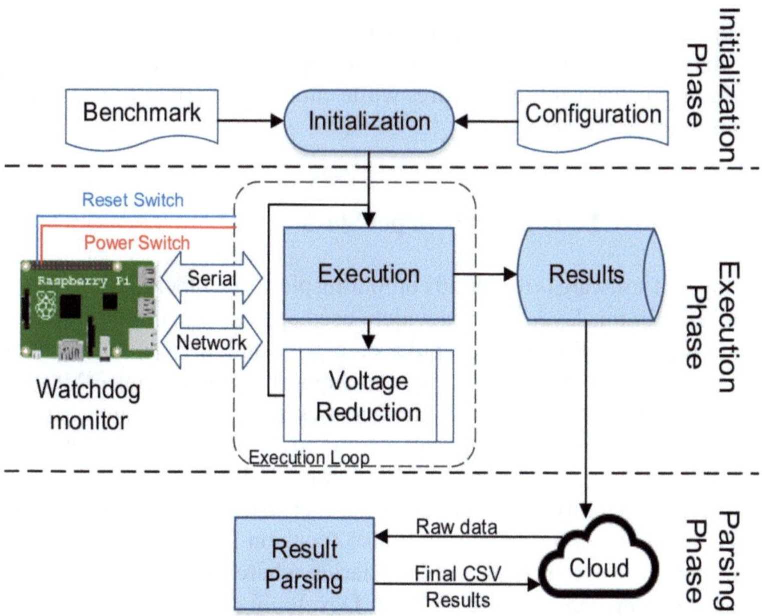

Fig. 13.3 Framework layout

- compares the outcome of the program with the correct output of the program when the system operates in nominal conditions to record Silent Data Corruptions (SDCs),
- monitors the exposed corrected and uncorrected errors from the hardware platform's error reporting mechanisms,
- recognizes when the system is unresponsive to restore it automatically,
- monitors system failures (crash reports, kernel hangs, etc.),
- determines the safe, unsafe, and non-operating voltage regions for each application for all frequencies, and
- performs massive repeated executions of the same configuration.

The automated framework (outlined in Fig. 13.3) is easily configurable by the user and can be embedded to any Linux-based system, with similar voltage and frequency regulation capabilities.

To completely automate the characterization process, and due to the frequent and unavoidable system crashes that occur when the system operates in reduced voltage levels, we set up a Raspberry Pi board connected externally to the X-Gene 2 board which behaves as a watchdog. The Raspberry is physically connected to both the Serial Port and the Power and Reset buttons of the system board to enable physical access to the system.

Below, we present the challenges that were taken into consideration for a solid development of such a framework.

Safe Data Collection. Given that a system operating beyond nominal conditions often has unexpected behaviors (e.g., file system driver failures), there is the need to correctly identify and store all the essential information in log files (to be subsequently parsed and analyzed). The automated framework was developed in such a way to collect and store safely all the necessary information about the experiments.

Failure Recognition. Another challenge is to recognize and distinguish the system and program crashes or hangs. This is a very important feature to easily identify and classify the final results, with the most possible distinct information concerning the characterization.

Reliable Cores Setup. This means that the cores, where the benchmark runs, must be isolated and unaffected from the other active processes of the kernel in order to capture only the effects of the desired benchmark.

Iterative Execution. The nondeterministic behavior of the characterization results due to several microarchitectural features making it necessary to repeat the experiments multiple times with the same configuration to eliminate the probability of misleading results.

In the following subsections, each of these functionalities grouped in the three distinct phases (i.e., Initialization, Execution, and Parsing) is discussed in detail.

13.6.1.1 Initialization Phase

During the *initialization phase*, the user can declare a benchmark list with any input dataset to run in any desirable characterization setup. The characterization setup includes the voltage and frequency (V/F) values under which the experiment will take place and the cores where the benchmark will be run; this can be an individual core, a pair of cores, or all of the available eight cores in the microprocessor.

This phase is in charge of setting the voltage and frequency ranges, the initial voltage and frequency values, with which the characterization begins, and to prepare the benchmarks—their required files, inputs, outputs, as well as the directory tree where the necessary logs will be stored. This phase is performed at the beginning of the characterization and each time the system is restored by the Raspberry (for example, after a system crash) in order to proceed to the next run until the entire Execution Phase finishes. Each time the system is restored, this phase restores the initial user's desired setup and recognizes where and when the characterization has been previously stopped.

The benchmark must run in an "as bare as possible" system without the interference of any other running process. Therefore, reliable cores setup is twofold—first, it recognizes these cores or group of cores that are not currently under characterization, and migrates all currently running processes (except for the benchmark) to a completely different core. Second, given that all the PMDs in the studied system are in the same power domain, they always have the same voltage value. On the other hand, each individual PMD can have different frequency, so all other cores are set to the minimum available frequency.

13.6.1.2 Execution Phase

After the initialization phase, the framework enters the Execution Phase, in which all runs take place. The runs are executed according to user's configuration, while the framework reduces the voltage with a step defined by the user in the initialization phase. For each run, the framework collects and stores the necessary logs at a safe place externally to the system under characterization, which will be then used by the parsing phase.

The logged information includes: the output of the benchmark at each execution, the corrected and uncorrected errors (if any) collected by the Linux EDAC Driver, as well as the errors' localization (L1 or L2 cache, DRAM, etc.), and several failures, such as benchmark crash, kernel hangs, and system unresponsiveness. The framework can distinguish these types of failures and keep logging about them to be parsed later by the parsing phase. Benchmark crashes can be distinguished by monitoring the benchmark's exit status. On the other hand, to identify the kernel hangs and system unresponsiveness, during this phase the framework notifies the Raspberry when the execution is about to start and also when the execution finishes.

In the meantime, the Raspberry starts pinging the system to check its responsiveness. If the Raspberry does not receive a completion notification (hang) in the given time (we defined as timeout condition a 2 times the normal execution time of the benchmark) or the X-Gene 2 turns completely unresponsive (ping is not responding), the Raspberry sends a signal to the Power Off button on the board, and the system resets. After that, the Raspberry is also responsible to check when the system is up again, and sends a signal to restart the experiments. These decisions contribute to the *Failure Recognition* challenge. We also ensure that any logging information will be stored correctly and no information will be lost or changed in case of any unstable system conditions (*Safe Data Collection*).

13.6.1.3 Parsing Phase

In the last step of our framework, all the log files that are stored during the Execution Phase are parsed in order to provide a fine-grained classification of the effects observed for each characterization run. Note that, each run is correlated to a specific benchmark and characterization setup. The categories that are used for our classification are summarized in Table 13.1, but the parser can be easily extended according to the user's needs. For instance, the parser can also report the exact location that the correctable errors occurred (e.g., the cache level, the memory, etc.) using the logging information provided by the Execution Phase.

Note that, each characterization run can manifest multiple effects. The characterization runs with the same configuration setup of different campaigns may also have different effects with different severity. For instance, let us assume two runs with the same characterization setup of two different campaigns. After the parsing, the first run finally revealed some CEs, and the second run was classified as SDC. To quantify the criticality of the effects of different experimental runs of different

Table 13.1 Experimental effect categorization

Effect	Description
NO (Normal operation)	The benchmark was successfully completed without any indications of failure
SDC (Silent data corruption)	The benchmark was successfully completed, but a mismatch between the program output and the correct output was observed
CE (Corrected error)	Errors were detected and corrected by the hardware
UE (Uncorrected error)	Errors were detected, but not corrected by the hardware.
AC (Application crash)	The application process was not terminated normally (the exit value of the process was different than zero)
SC (System crash)	The system was unresponsive; meaning that the X-Gene 2 is not responding to pings or the timeout limit was reached

Table 13.2 Weights used in our experiments

Weight	Value
W_{SC}	16
W_{AC}	8
W_{SDC}	4
W_{UE}	2
W_{CE}	1
W_{NO}	0

campaigns with the same setup, we define the "severity function" S_v, where v is the voltage value, as presented below:

$$S_v = W_{SDC} \cdot SDC/N + W_{CE}CE/N + W_{UE} \cdot UE/N + W_{AC} \cdot AC/N + W_{SC} \cdot SC/N$$

In this function, the parameters *SDC*, *CE*, *UE*, *AC* and *SC* can take the values from 0 to *N* (*N* is the number of runs at each voltage level), and represent the times that this effect appears to these runs. Parameters W_{SDC}, W_{CE}, W_{UE}, W_{AC} and W_{SC} represent "weights" that can be set to characterize the severity of each effect of Table 13.1. The higher the weight, the more critical the effect is considered by our function. For the purpose of this chapter, we defined the values presented in Table 13.2 as values for our severity function (any values for the weights can be used).

In this experimental research, two different benchmarks are used: *Linpack*, which is a widely-used high-performance benchmark [35] and *hmmer* from the SPEC CPU2006 benchmark suite [36]. Both of them ran on a single core, while the remain-

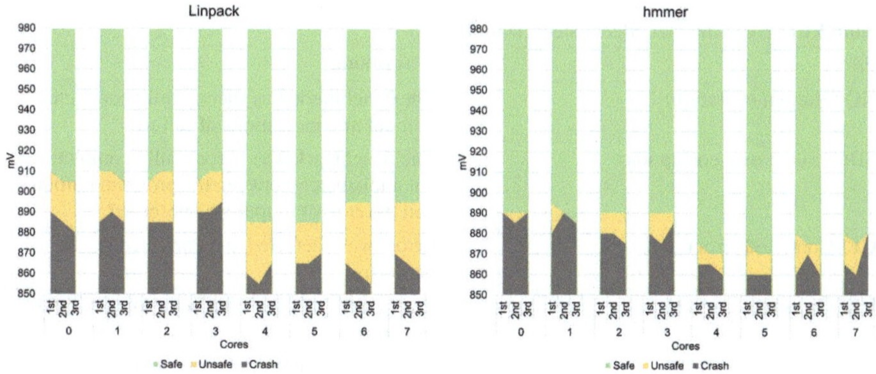

Fig. 13.4 Cores characterization

ing cores are reliable and each campaign were ran three times. Figure 13.4 presents the three campaigns for the Linpack and hmmer benchmark (1st, 2nd, 3rd), respectively, in the case that are executed in each individual core running at 2.4 GHz, while the rest of the cores operate on the reliable cores setup. In both benchmarks, we can notice the three regions of operation according to the collected results. The regions are given as follows:

- *Safe region* (green): NO (normal operation) without SDCs, errors or crashes.
- *Unsafe region* (yellow): abnormal behavior (SDC, CE, UE, AC) but not a system crash.
- *Crash region* (grey): SC (system crash).

It is clear that there is a significant variation among the three runs and significant core-to-core static variation for the same benchmark. From these results, it follows that cores 4 and 5 of the particular chip were used more robust than the others. Moreover, it is important to notice the width of the *Safe* region in the two benchmarks that is up to 11.2% lower than the nominal voltage value (980 mV). This reduction of the voltage from the nominal value corresponds to power gains up to more than 21%. Finally, the width of the *Safe* region ranges from 0 mV up to 40 mV. This illustrates that with the development of appropriate mitigation techniques the power gain can reach the 28.3%.

Finally, for the same characterization runs we used the severity function before to present the scaling of the effects and their severity in the reduced voltage margins. In Fig. 13.5, we can notice that the lighter the color, the more stable and reliable is the system. While reducing the voltage margins, we observe that the instability increases (the color becomes darker), until the darkest color, which indicates that the system cannot operate in such voltage margins.

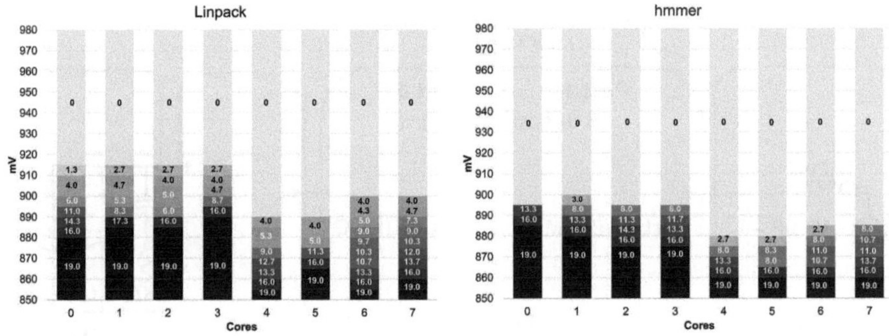

Fig. 13.5 Severity scaling

13.6.2 *IR-Drop Analysis*

As we discussed, a significant percentage of the margins adopted in voltage or frequency is attributed to dynamic variations such as IR-drop, thus a significant effort in UniServer is spent on evaluating it. In particular, a two-pronged strategy for direct analysis and measurement of power supply noise is developed. For the purposes of analysis, the ARM Juno platform was chosen for several compelling reasons. First, the Juno platform integrates a high-performance (excess of 1.2 GHz) dual-core ARM Cortex-A57 cluster. Second, the Juno platform is a dual-core Cortex-A57 cluster. System-architecturally, this is very similar to other high-performance multi-core ARM systems. Thus, the software interactions that we can reveal in our Juno analysis should also hold true for these high-performance platforms as well. We particularly highlight the impact of multi-core execution and power gating on supply voltage noise.

The third compelling reason for the choice of the Juno platform is the availability of an on-chip Digital Storage Oscilloscope (OC-DSO) [10] that was designed and integrated for direct snooping of on-chip voltage noise on the Cortex-A57 cluster. Our knowledge of this voltage monitoring peripheral enables us to develop a comprehensive analysis framework.

In the following subsection, we describe our analysis setup for the Juno platform and some of our measurement results.

13.6.2.1 Juno PDN Simulation Framework

We designed and developed a simulation methodology to analyze the impact of supply noise. Our simulation methodology enables us to execute industrial workloads which we then compare against direct measurement results to validate the model.

Figure 13.6 illustrates the power delivery network model [37]. The hybrid model incorporates a combination of lumped elements and distributed network models to

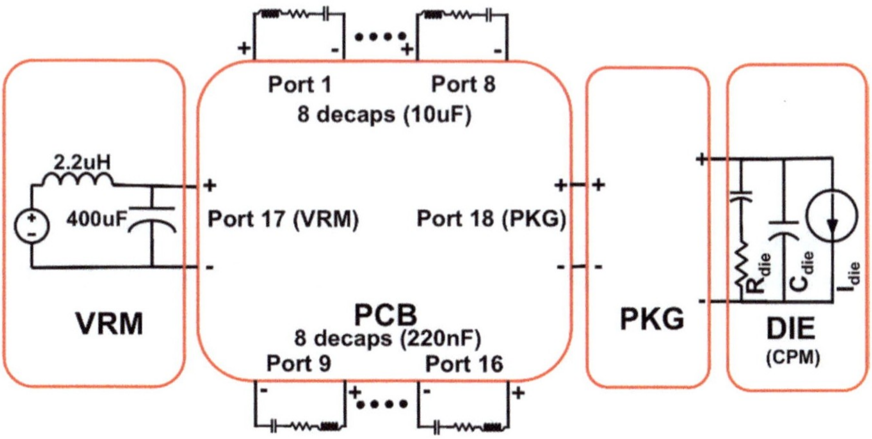

Fig. 13.6 PDN simulation model consisting of lumped and distributed parameters

efficiently model system behavior across a wide frequency range. Accurate power delivery network (PDN) behavior at low-frequency requires a closed-loop, small-signal model of the VRM, with a voltage-feedback sense-point, and an appropriate feedback-compensation model. Instead, we represent the VRM by an open-loop, small-signal lumped circuit model. Such an approach retains PDN accuracy at the mid- and high-frequency ranges (100 kHz–100 MHz) without increasing overall simulation and modeling complexity.

We extract a lumped chip power model (CPM) for the A57 compute cluster using Apache Redhawk [38]. The lumped model of the die consists of a current source that represents switching transistors. Non-switching transistors act as local decoupling capacitors that provide instantaneous current demands. The power grid resistance (Rdie) is also modeled in the power model for the die. Resonance interactions between the die and the rest of the PDN network occur at frequencies in the range around 100 MHz, where a lumped circuit model is sufficient for accurate modeling of the die.

13.6.2.2 PDN Simulation Results

Figure 13.7 shows the input impedance of the PDN as a function of frequency. The impedance is represented in the db-Ohm scale ($1\ \Omega = 0$ dB) in order to highlight key attributes of the PDN that may not be seen on the linear scale. The dB scale helps to highlight the *relative* difference in magnitude, i.e., every 6 dB reduction is equivalent to a magnitude difference of 50%.

The VRM (voltage regulator module), PCB (printed circuit board) decoupling capacitors and the die capacitance form three parallel impedance branches of the overall PDN. At low frequencies (<1 kHz), the path of least impedance is through the VRM. The series inductor (2.2 μH) in the switching regulator circuit dominates

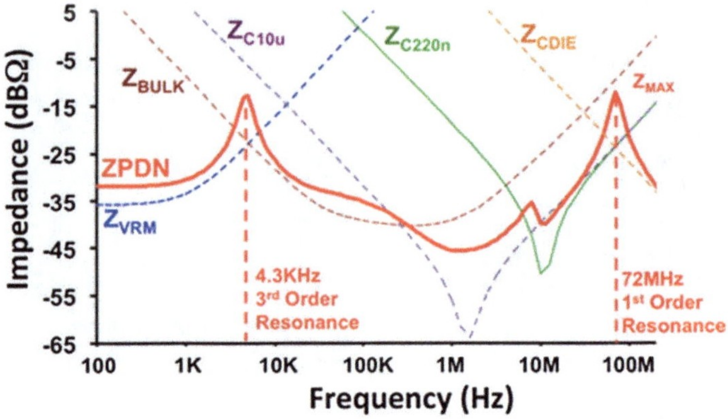

Fig. 13.7 PDN frequency domain simulation results

overall VRM impedance (Z_{VRM}). The inductor and the bulk-capacitor (400 μF) at the VRM output form a LC-tank circuit that resonates at 4.5 kHz. This represents the *third-order* resonance frequency of the system PDN.

The impedance of the VRM bulk-capacitor (Z_{BULK}) is ultimately limited by its ESL, which causes inductive behavior beyond the self-resonance frequency of 750 kHz. The PCB decap network consists of eight capacitors of value 10 μF each and an additional eight capacitors of value 220 nF. The set of 10 μF capacitors have a self-resonance frequency of 1 MHz. Beyond this, their frequency response is dominated by the parasitic inductance loop formed by the series connection of the decap ESL and the PCB trace inductance connecting the capacitor to the die bumps. Looking in from the decap pads as a single lumped port, this parasitic loop inductance was measured to be 143 pH (in simulation) for the set of eight 10 μF capacitors.

The parasitic inductance of the PCB decaps is dominated by their connection to the PCB. The ESL of individual capacitors is ~400 pH, whereas the PCB connections from the decaps to the package add an effective inductance of 1.2 μH. Thus, the ESL of each capacitor element affects the overall inductance weakly, which is dominated as such by PCB parasitics.

13.6.2.3 Measurement Results on the Juno PDN

Figure 13.8 shows our on-chip measurement framework. A high-bandwidth on-chip digital storage oscilloscope (OC-DSO) snoops the supply rails of the Cortex-A57 cluster. The OC-DSO runs continuously in real-time, logging data and capturing waveforms on trigger events. Event counter and tide-mark registers track the size and frequency of voltage transients. For voltage transients of interest, threshold and gradient triggers can initiate waveform capture of up to 2 K points into the internal SRAM trace buffer. A decimation block allows flexible adaptation of the bandwidth/sample rate to allow measurement of low-frequency transients.

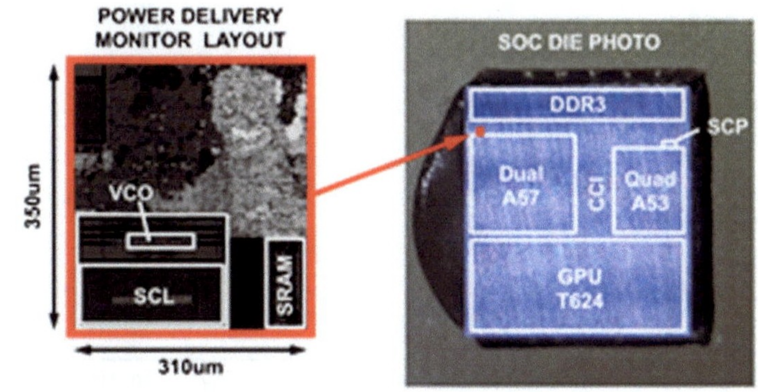

Fig. 13.8 PDN measurement setup—the on-chip digital storage oscilloscope (OC-DSO) design and integration

The on-chip measurement setup enables high-bandwidth probing of the internal supply rails of the A57. Accuracy of the time-domain measurement is limited by the minimum resolution of the internal ADC and the internal noise generated by the clocked transistors. Therefore, the minimum voltage droop that can be measured using this technique is in the range between 5 and 10 mV. The limitations on the maximum current draw of the SCL block and the accuracy of the internal ADC limit the minimum impedance magnitudes ($\sim$50 m Ω) that can be measured using this technique.

13.6.2.4 Worst-Case Resonant Code Generation

Manually creating workloads that can trigger worst-case resonances in the system is difficult due to the complexity of the underlying microarchitecture, especially in out-of-order cores, such as the ARM A57. This issue was circumvented by automatically generating worst-case workloads using a genetic-algorithm-based framework that is agnostic to the processor microarchitecture.

In this experimental study, the genetic algorithm setup outlined in Sect. 13.6.1.2. to develop resonant workloads for the Juno platform was used. An initial seed population of instructions is generated from the packaged vectors delivered with the A57 processor IP. The algorithm uses voltage noise measurements, from the on-chip oscilloscope circuitry as the optimization objective function. Selection criteria, based on the droop magnitude, prune the workload population by selecting the ones causing the maximum voltage droops. These serve as "parents" which are then paired and mutated to create the next generation of workloads. The process is iterated until the increase in voltage droop across succeeding generations saturates. The fittest instruction in the final iteration is chosen as the worst-case workload (the "GA_Max_Droop" workload). This approach allows flexibility since different measurement variables (such as average current consumption) can also be chosen as the optimization objectives.

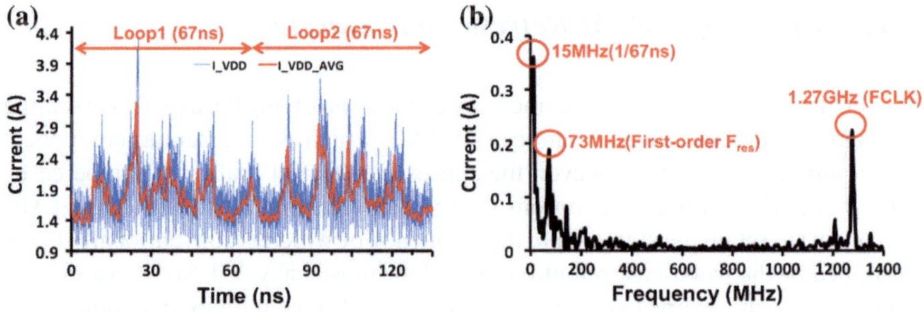

Fig. 13.9 Simulated current waveform (I_VDD) for two successive loops of the "GA_Max_Droop" workload

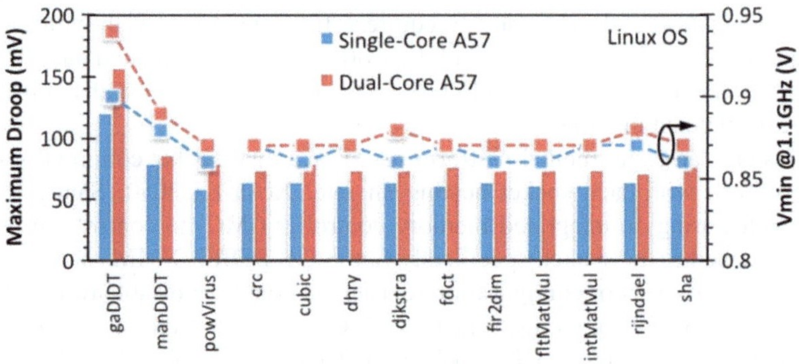

Fig. 13.10 Vmin and maximum voltage noise for various workloads

Genetic-algorithm-based approaches often converge on local minima and may not generate the global minima. Therefore, the output of the algorithm is evaluated in terms of its efficacy in exciting the worst-case resonance. Figure 13.9 shows the simulated current waveform (I_VDD) for two successive loops of the "GA_Max_Droop" workload. The current waveform appears discontinuous due to an ideal PDN being used, with infinite current bandwidth. Each loop is 67 ns in duration when the processor is operated at the clock frequency of 1.27 GHz. This is close to the maximum frequency of operation for the compute cluster, as measured in silicon.

Figure 13.10 shows how the "gaDIDT" workload (the GA generated voltage noise virus) compares to other workloads in terms of voltage noise and Vmin (minimum operational voltage for a given frequency). Measurements from single- and dual-core, the "gaDIDT" workload (the GA-generated voltage noise virus) compares to other workloads in terms of voltage noise and Vmin (minimum operational voltage for a given frequency). Measurements from single- and dual-core runs are shown. On dual-core runs both voltage noise and Vmin are higher. The voltage noise virus generates 2× more voltage droop than conventional workloads. Also, the Vmin measurements suggest that the operational voltage for 1.1 GHz frequency can be reduced by at least 6% for dual-core operation and by 10% for single-core operation.

13.6.3 Revealing DRAM Refresh-Rate Margins

The ever-increasing need for higher memory capacity is driving the aggressive scaling of Dynamic Random-Access Memory (DRAM), which is an essential component of all computing systems. However, the aggressive DRAM scaling is hampered by the need of periodic refresh operations (triggered by the default Auto-Refresh (AR) mechanism) to retain the stored data, the frequency of which is conventionally being determined by the worst-case retention time of the most leaky cell. Such an approach might help to achieve error-free storage but its viability is in doubt due to the large waste of power and throughput that may incur reaching up to 25–40% and 15–30% respectively, in future 32–64 GB densities [11, 12]. To address such an alarming challenge, many recent studies have shown that the retention time of DRAM cells varies a lot and most of the cells do not require as frequent refresh as the conventional paradigm dictates [39]. The majority of the works have been evaluated on simulators or on experimental setups based on FPGAs and not in a real server with a full system software as planned in UniServer.

To this end, an experimental platform based on a dual-socket commodity server was developed with each socket hosting an Intel Xeon E5-2650 (Sandy Bridge) processor featuring an integrated memory controller (iMC) to control the DRAM devices attached to the socket, specifically four 8 GB DDR3 DIMMs at 1600 MHz. The iMC exposes a set of configuration registers to enable or disable refresh for the entire DRAM. Following the fact that each iMC controls a single memory domain, DRAMs attached to different CPU sockets can have the auto-refresh enabled or disabled separately. In the developed dual-socket system depicted in Fig. 13.11, the first memory domain is deemed reliable with the nominal refresh-rate applied, in which

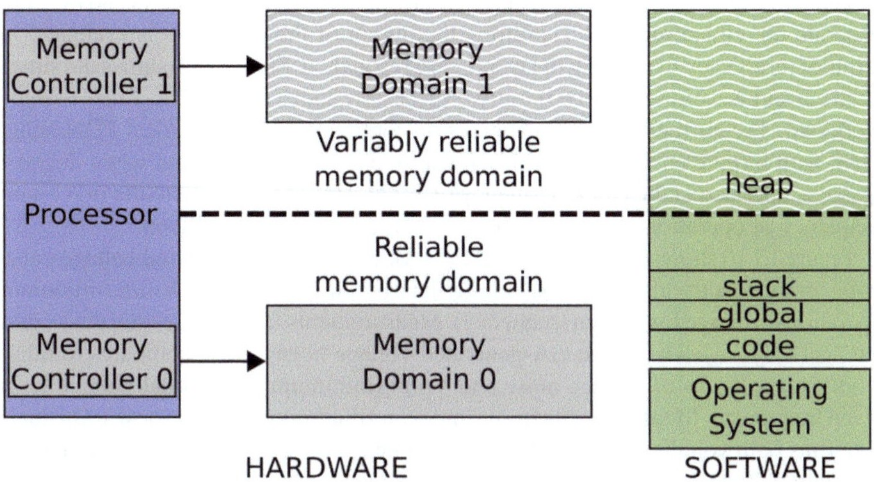

Fig. 13.11 Experimental setup with different memory reliability domains

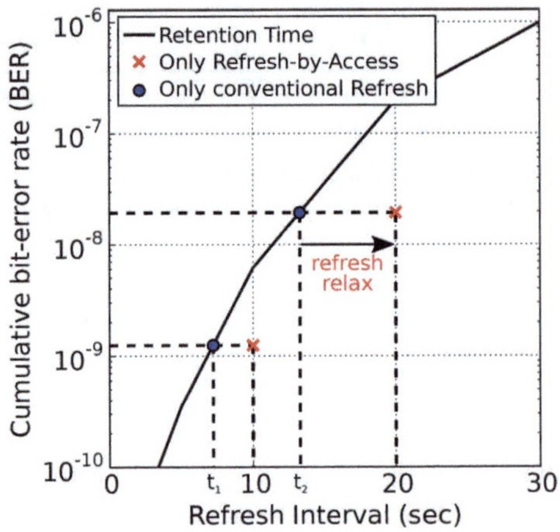

Fig. 13.12 Cumulative distribution function of bit-errors for relaxed refresh rate

the Operating System is allocated, while the second one is configured with relaxed or no refresh. Such configuration can safeguard the kernel data that cannot tolerate errors in the reliable domain for avoiding any catastrophic failures and execute the application on the second memory domain [40].

Utilizing such an experimental setup in server room temperature (20–23 °C) and using known memory test patterns [41], the Bit-Error Rate (BER) of 8 GB DIMMs have been characterized by aggressively relaxing the refresh-rate from the conservative 64 ms to 1 s and up to 30 s. The resulted Cumulative Distribution Function (CDF) is depicted in Fig. 13.12 from which it is evident that as the refresh rate is being relaxed the BER is increasing. An interesting observation is that the BER is maintained below 10^{-10} even for a refresh-rate less than 5 s, which is almost 78 times less than the conventional refresh rate.

Apart from the rather rare and extremely pessimistic events and conditions that such experiments may reveal another factor that could be exploited for further extending the refresh while potentially minimizing the BER are the implicit refreshes that take place during each access. In fact, every DRAM access naturally opens the accessed row and consequently restores the leaked charge in the capacitor of DRAM cells, thus incurring an implicit refresh operation. Refresh-by-Access (RefA) can be exploited to significantly relax the refresh rate while restricting the number of manifested errors (see Fig. 13.12).

To understand such a concept, it is worth to consider a simple scenario as depicted in Fig. 13.13 and assume that a running application triggers N_r Memory Accesses (MemA) to the r address in memory so that

$$MemA_r = \{MemA_{0,r}, \ldots, MemA_{Nr,r}\}$$

Clearly, the intervals between consecutive accesses to the same address can be calculated as:

$$\Delta t_{i,r} = t_{MemAi+1,r} - t_{MemAi,r}$$

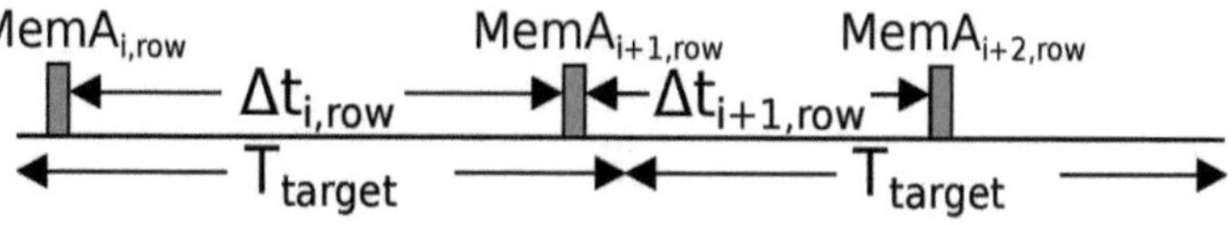

Fig. 13.13 Graphical representation of our approach where all time intervals between consecutive memory accesses are being kept lower than a target time interval (target retention time)

If the maximum of all $\Delta t_{i,r}$ is smaller than a target retention time T_{target} of the DRAM cells, i.e.,

$$\max(\Delta t_{i,r}) \leq T_{target}$$

then, all cells would be implicitly refreshed though the memory accesses. In this case, the conventional AR can be relaxed up to T_{target} while maintaining the actual BER low, since no cell that has up to T_{target} retention time will fail. In fact, the BER will be bounded to a value lower than when we adopt AR of T_{target}

$$BER(max(\Delta t_{i,r})) \leq BER_{target}$$

To showcase such an approach, an artificial benchmark was developed such that it issues as many memory accesses as required for touching all the stored data iteratively within intervals of 10–20 s. By doing so, it was essentially ensured that $\max(\Delta ti, r) = \{10, 20\}$ s in each case. The results are shown in Fig. 13.12 as crosses. An interesting observation is that by ensuring that all memory access intervals are bound to a value, then the resulting BER is equal to the one achieved by using AR with a more frequent refresh. In particular, the first case resulted in a BER of 10^{-9} that is equal to the one achieved with *AR* of 8.6 s while the second case resulted in a BER of 2×10^{-8} that is equivalent to AR of 12 s. This means that we can relax the refresh by 16% or by 66% or even omit the *AF* and achieve the same BER.

13.6.3.1 Application to Stencil Algorithms

To evaluate the effectiveness of the above approach on real applications, the above scheme to stencil algorithms, which are very popular in various high-performance application, has been applied. Stencil-based algorithms are a class of iterative kernels. In each sweep, the stencil updates all the elements of a n-dimensional grid using neighboring elements in a fixed pattern called stencil. The naive implementation for the stencil is by using nested loops for computing each sweep sequentially. Looping implementation lack from poor cache performance.

We are using Pochoir stencil compiler, which is built on top of Cilk Plus multi-thread extension and uses trapezoidal decompositions, which utilize the cache more efficiently. The decomposition is breaking down the problem into smaller tasks that are responsible for a part of the grid and the associated stencil for a number of

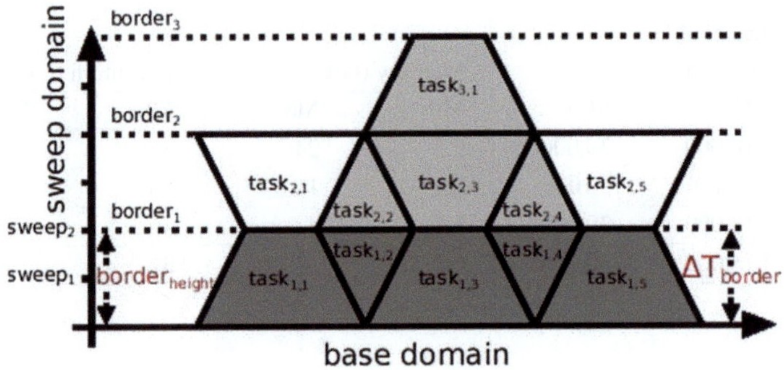

Fig. 13.14 Trapezoidal decomposition of 1D problem. All the tasks below border1 must be completed in order to continue to next sweep

sweeps. It decomposes the sweep domain when the number of sweeps is greater than a threshold and recursively process the lower sub-trapezoids before the upper ones or the base domain so that it ends up fitting to the size of the cache. The compiler was modified to schedule the memory accesses to facilitate RefA. For the proposed method, the interval between consecutive memory accesses should be controlled. For this reason, borders are introduced. Every border height sweeps and breaking down the sweep domain on the same sweep for all the grid, as shown in Fig. 13.14 with the dashed line. The border essentially does not allow to start any consequent sweep without completing all the tasks in the current border height. We can ensure that each element in the grid will be accessed during that parameter set of p = {p1, p2, …} that determine accesses $\{MemA_{i,r}\}$.

The time intervals between consecutive borders can be measured. As all elements are accessed at least once in each $border_{height}$, each $\Delta t_{i,r}$ is bound to be less than the duration of computing all the tasks in the current $border_{height}$, which is denoted as interval Δt_{border}. As shown previously, the $max(\Delta t_{i,r})$ will is than Δt_{border}. Figure 13.14 shows also the *Original* implementation that would opportunistically progress as many sweeps as possible, specifically after finishing $task_{1;1}$, $task_{1;2}$, and $task_{1;3}$, it will compute all the tasks up to the $task_{3;1}$, highlighted in grey.

In this experiment, $border_{height}$ is set as a parameter to control the Δt_{border}. Lowering the $border_{height}$ will reduce the required computation of each task and the Δt_{border}. The $border_{height}$ must be carefully selected so that the BER resulted from the Δt_{border} will be kept under the set threshold, BER_{target}. Furthermore, tasks should not be broken down to very small ones as there is a considerable performance overhead caused by the function calls and by breaking the cache efficiency.

The order of the tasks which belong to the same border height is scheduled in such a way as to ensure that tasks which are responsible for the same area of the grid run consecutively and achieve the maximum performance efficiently under the same constraints.

To evaluate the proposed approach, the Pochoir compiler was used for a range of stencil-based algorithm which has different number of dimensions, grid size, data size

Table 13.3 Benchmarks

Benchmark	Grid size		Memory (GB)		Execution time (s)	
	Min	Max	Min	Max	Min	Max
Head-2D	18,000	30,000	9.3	24	38	96
Head-3D	780	1020	7.4	16.6	56	263
Life-2D	18,000	30,000	1.5	4	46	156
Artf-2D	18,000	30,000	6.2	16	37	98

of each element, and complexity of the stencil. Specifically, we have experimented with the algorithms of Heat Dissipation (*Heat*) [17], Conways's Game of Life (*Life*) [17] and an artificial benchmark (*Artf*), the characteristics of which are shown in Table 13.3. The *Artf* is designed as a 3×3 kernel on two-dimensional grid which in each *sweep* applies binary arithmetic and shifting between different elements so that each bit-error will persist during the execution and be manifested once at the end result.

The chosen algorithms are not resilient to errors, and thus even one error can have catastrophic results since this error may propagate to neighboring elements. The quality of the measured metric is estimated as the percentage of runs that completed correctly. For further characterization and to be able to measure the number of bit-errors, the *Artf* is used.

13.6.3.2 Experimental Results

(1) *Artificial Benchmark*: We start with the results obtained from *Artf* as it is possible to measure manifested errors. We are comparing the benchmark compiled with the Original implementation of the Pochoir and our Proposed one. In our experiments, we sweep the border_{height} from 10 to 40 and the grid size from 18,000 to 30,000. The border_{height} in the results of Fig. 13.6 is chosen to be 20 so that the following constraints are met: (i) the maximum introduced performance overhead is lower than 15% and (ii) the maximum Δt_{border} is below 2 s that corresponds to BER of 10^{-10}. The results are obtained after executing *Artf* 1340 times while varying the grid size. The left plot of Fig. 13.15 compares the percentage of the correct runs of the two implementations and the right plot of Fig. 13.6 shows the average number of bit-errors occurred. We can observe that given the constraints, our *Proposed* implementation outperforms retaining a high percentage, over 95%, of correct runs across the range of grid sizes, while the Original implementation rapidly decrease its quality.

(2) *Application Benchmarks*: Figs. 13.16 and 13.17 present the design space exploration for the performance overhead and for the Δt_{border} of the benchmarks Heat-2D and Heat-3D. The design space is populated from the results of runs with variable border_{height} and grid size. We are sweeping the border_{height} for Heat-2D

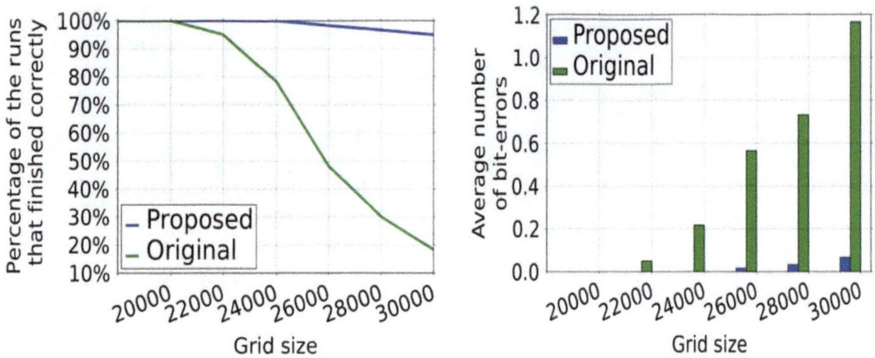

Fig. 13.15 Errors characterization for Artf benchmark. The left plot—percentage of the runs that completed correctly; the right plot—average number of errors per run

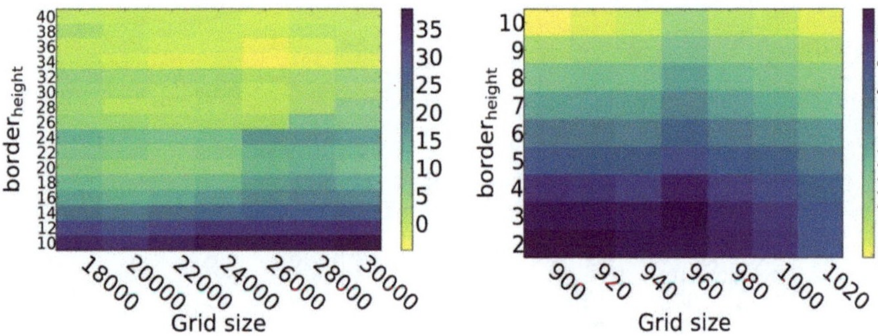

Fig. 13.16 Percentage of performance overhead introduced from our technique

from 10 to 40 and for Heat-3D from 2 to 10. Respectively, we are sweeping the grid size from 18,000 to 30,000 and from 900 to 1020. We observe that the performance overhead is mainly affected by the border_{height} caused by the disruption of cache locality and by the increased number of function calls while having smaller tasks. Δt_{border} is affected by both the border_{height} and the grid size, as the border_{height} affects the number of stencil computations for each task and the grid size affects the number of tasks that belong in each border_{height}. We are using the design space to select the parameter values that minimize refresh operations given different constraints such as maximum acceptable performance overhead and the maximum Δt_{border}.

Based on the performed design space exploration, we analyze three possible scenarios for better understanding the efficacy of our approach. We are selecting parameters to optimize: (i) for the minimum Δt_{border} and consequently the lowest BER, (ii) for the least performance overhead and (iii) for both those two previous criterion together. So we are exploring the error characteristics for the three scenarios with $\text{border}_{height} = \{10, 40, 20\}$, equivalent $\max(\Delta t_{border}) = \{1.5, 4.5, 2.4\}$ seconds and

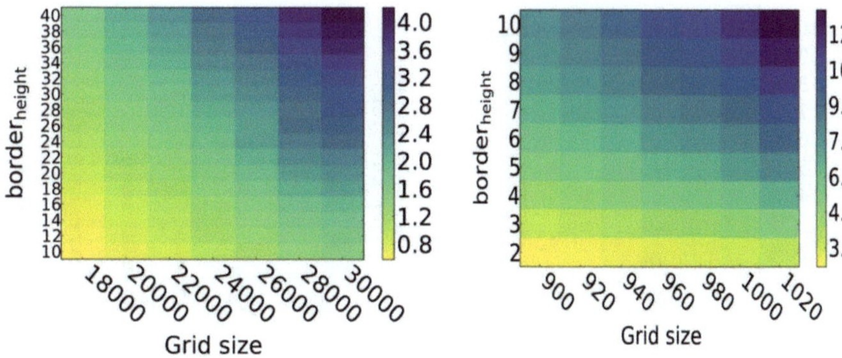

Fig. 13.17 Achieved Δt_{border} in seconds

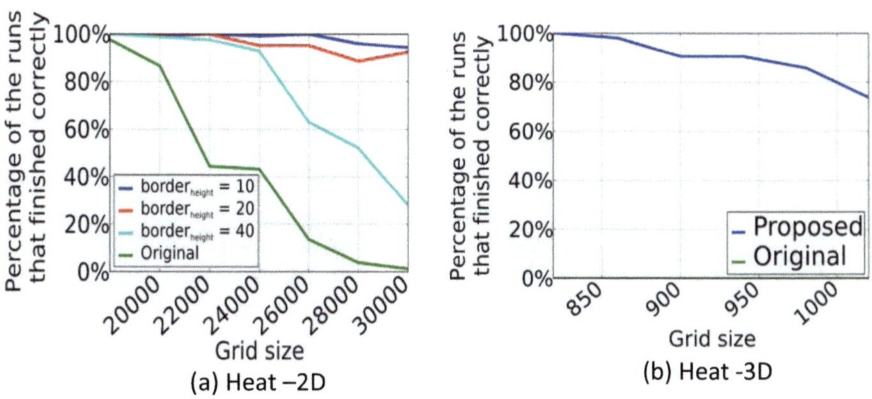

Fig. 13.18 Percentage of the runs that completed correctly

max(overhead) = {38%, 5%, 12%}. We can see the results of 1968 runs in Fig. 13.18a, which depicts the percentage of the runs that completed correctly.

We can observe that for the first and third scenario, the percentage of correct runs remain over 90% for all the grid sizes, while for border$_{height}$ = 40, the percentage decreases more rapidly. However, in each case, the results are better than the Original implementation.

In Fig. 13.18b, we explore one setting for Heat-3D with parameter of border$_{height}$ = 6 with equivalent max(Δt_{border}) = 8 s and max(overhead) = 44%.

For the selected range of grid sizes, the original implementation has no devastating results even in the smaller grids, while the quality degrades slowly with our Proposed implementation.

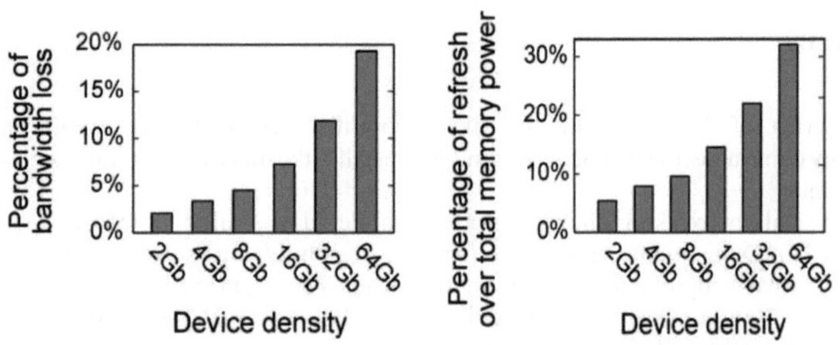

Fig. 13.19 Refresh implications on power and performance on current and future DRAM technologies [1]. The left plot depicts throughput loss due to refresh. The right plot shows power overhead of refresh operations

13.6.3.3 Performance and Power Gains

Based on the above results, it is evident that there are cases where our approach can relax or omit refresh with minor quality losses. If we omit the AR, we obviously help to limit the AR overheads. As seen in Fig. 13.19, we can gain from 4.4% performance in current technologies and up to 19% in future technologies and decrease the power consumption of the memory system by 9.6–31%.

13.7 Conclusion

This Chapter discussed the variability challenges and their impact on energy efficiency of servers. The basic ideas of the UniServer [43] paradigm were presented which attempts to reduce hardware safety margins by utilizing representative stress cases, constant hardware monitoring and predictive mechanisms. The complete system stack approach includes a modified error-resilient Hypervisor and a cloud resource management software. The results of the initial phase of the project indicate the possible margins existing in the state-of-the art CPUs and DRAMs while revealing the margins attributed to IR-drop. Efficient schemes are needed to ensure disruptive operation by bullet-proofing the system software the development of which is the aim of the second phase of the project. The developed technologies are integrated within a 64-bit ARM-based microserver which aspires to drive Edge computing and turn the opportunities in the emerging Big Data and IoT markets into smarter products.

References

1. The Internet of Things: sizing up the opportunity (2014) Online report. http://www.mckinsey.com/industries/semiconductors/our-insights/the-internet-of-things-sizing-up-the-opportunity
2. Cisco visual networking index: global mobile data traffic forecast update 2016–2021 (2017). Online white paper. http://www.cisco.com/c/en/us/solutions/collateral/service-provider/visual-networking-index-vni/mobile-white-paper-c11-520862.html
3. Hamilton J (2009) Cooperative expendable micro-slice servers (CEMS): low cost, low power servers for internet-scale services. IEEE, Asilomar
4. Shi W, Cao J, Zhang Q, Li Y, Xu L (2016) Edge computing: vision and challenges. IEEE Internet Things J 3(5):637–646 2016
5. Koomey JG et al (2011) Implications of historical trends in the electrical efficiency of computing. IEEE Ann Hist Comput 33(3):46–54
6. Esmaeilzadeh H et al (2011) Dark silicon and the end of multicore scaling. In: Proceedings of the 38th IEEE annual international symposium on computer architecture (ISCA), pp 365–376
7. Ghosh S, Roy K (2010) parameter variation tolerance and error resiliency: new design paradigm for the nanoscale era. Proc IEEE 98(10):1718–1751
8. Karakonstantis G, Roy K (2010) Low power and variation-tolerant application-specific system design. In: Low-power variation-tolerant design in nanometer silicon. Springer
9. Esmaeilzadeh H et al (2011) Dark silicon and the end of multicore scaling In: IEEE international symposium on computer architecture (ISCA)
10. Whatmough PN et al (2015) 14.6 an all-digital power-delivery monitor for analysis of a 28 nm dual-core arm cortex-a57 cluster. In: IEEE ISSCC
11. Jamie L et al (2013) An experimental study of data retention behavior in modern DRAM devices. In: IEEE ISCA '13
12. Qureshi MK et al (2015) Avatar: a variable-retention-time (VRT) aware refresh for dram systems. In: Proceedings of the 2015 45th IEEE DSN '15, pp 427–437
13. Borkar S et al (2003) Parameter variations and impact on circuits and microarchitecture. In: IEEE DAC
14. Reddi VJ et al (2010) Voltage smoothing: characterizing and mitigating voltage noise in production processors via software-guided thread scheduling. In: IEEE MICRO
15. Bacha A, Teodorescu R (2013) Dynamic reduction of voltage margins by leveraging on-chip ECC in itanium II processors. In: Proceedings of the 40th IEEE annual international symposium on computer architecture (ISCA), pp 297–307
16. Karakonstantis G et al (2001) Containing the nanometer pandora-box: crosslayer design techniques for variation aware low power systems. In: IEEE JETCAS
17. Leem L et al (2010) Cross-layer error resilience for robust systems. In: Proceedings of IEEE/ACM ICCAD
18. Das S et al (2009) RazorII: in situ error detection and correction for PVT and SER tolerance. In: IEEE JSSCC
19. Bowman K et al (2011) A 45 nm resilient microprocessor core for dynamic variation tolerance. IEEE J Solid-State Circuits 46(1)
20. Bull DM et al (2011) A power-efficient 32 bit ARM processor using timing error detection and correction for transient-error tolerance and adaptation to PVT variation. In: IEEE JSSC
21. Xu Q Kim NS, Mytkowicz T (2016) Approximate computing: a survey. IEEE Des Test
22. Mitra S et al (2011) Robust system design to overcome CMOS reliability challenges. IEEE J Emerg Sel Top Circuits Syst 1(1)
23. Sampson A, Dietl W, Fortuna E, Gnanapragasam D, Ceze L, Grossman D (2011) EnerJ: approximate data types for safe and general low-power computation. In Proceedings of the 32nd ACM SIGPLAN conference on programming language design and implementation (PLDI), New York, NY, June 2011
24. Esmaeilzadeh H, Sampson A, Ceze L, Burger D (2012) Architecture support for disciplined approximate programming. In: ASPLOS 2012, pp 301–312, London, UK

25. Cho H, Leem L, Mitra S (2012) ERSA: error resilient system architecture for probabilistic applications. IEEE Trans CAD Integr Circuits Syst 31:546–558
26. Narayanan S, Sartori J, Kumar R, Jones DL (2010) Scalable stochastic processors. In: DATE
27. Ganapathy S, Karakonstantis G, Teman A, Burg A (2015) Mitigating the impact of faults in unreliable memories for error-resilient applications. In: IEEE DAC
28. Teman A, Karakonstantis G et al (2015) Energy versus data integrity trade-offs in embedded high-density logic compatible dynamic memories. In: IEEE DATE
29. Gu et al W (2003) Characterization of linux kernel behavior under errors. In: IEEE DSN
30. David FM et al (2007) Building a self-healing operating system. In: IEEE DASC
31. Jin X et al (2015) FTXen: making hypervisor resilient to hardware faults on relaxed cores. In: IEEE HPCA
32. Bahga A et al (2012) Analyzing massive machine maintenance data in a computing cloud. In: IEEE TPDS
33. Dean D et al (2012) UBL: unsupervised behavior learning for predicting performance anomalies in virtualized cloud systems. In: ICAC
34. Gaikwad P et al (2016) Anomaly detection for scientific workflow applications on networked clouds. In: HPCS
35. Huzum C et al (2011) March test for static neighborhood pattern-sensitive faults in random-access memories. Elektronika ir Elektrotechnika
36. Dongarra JJ, Luszczek P, Petitet A (2003) The LINPACK benchmark: past, present and future. Concurr Comput pract Exp 15(9):803–820
37. Henning JL (2006) SPEC CPU2006 benchmark descriptions. SIGARCH Comput Archit
38. Das S et al (2015) Modelling and analysis of the system-level power delivery network for a dual-core cortex-A57 in 28 nm CMOS. In: International symposium on low-power electronic design (ISLPED)
39. Qiang G et al (2012) A failure detection and prediction mechanism for enhancing dependability of data centers. J Comput Theory Eng
40. Liu J et al (2012) RAIDR: retention-aware intelligent dram refresh. In: IEEE ISCA, pp 1–12
41. Tovletoglou K, Nikolopoulos D, Karakonstantis G (2017) Relaxing DRAM refresh-rate through access pattern scheduling. IEEE IOLTS
42. Karakonstantis G et al (2018) An energy-efficient and error-resilient server ecosystem exceeding conservative scaling limits. IEEE Des Test Eur 2018:1099–1104
43. Karakonstantis G, Nikolopoulos DS, Gizopoulos D, Trancoso P, Sazeides Y, Antonopoulos CD, Venugopal D, Das S (2017) Error-resilient server ecosystems for edge and cloud datacenters. IEEE Comput 50(12):78–81
44. Papadimitriou G, Kaliorakis M, Chatzidimitriou A, Magdalinos C, Gizopoulos D (2017) Voltage margins identification on commercial x86-64 multicore microprocessors. IEEE International Symposium on On-Line Testing and Robust System Design (IOLTS 2017), Thessaloniki, Greece, July 2017
45. Papadimitriou G, Kaliorakis M, Chatzidimitriou A, Gizopoulos D, Favor G, Sankaran K, Das S (2017) A system-level voltage/frequency scaling characterization framework for multicore CPUs. IEEE Silicon Errors in Logic – System Effects (SELSE 2017), Boston, MA, USA, March 2017
46. Papadimitriou G, Kaliorakis M, Chatzidimitriou A, Gizopoulos D, Lawthers P, Das S (2017) Harnessing voltage margins for energy efficiency in multicore CPUs. IEEE/ACM International Symposium on Microarchitecture (MICRO 2017), Cambridge, MA, USA, October 2017
47. Chatzidimitriou A, Papadimitriou G, Gizopoulos G (2018) HealthLog Monitor: A Flexible System-Monitoring Linux Service. IEEE International Symposium on On-Line Testing and Robust System Design (IOLTS 2018), Costa Brava, Spain, July 2018

Index

C. Kachris et al. (eds.), *Hardware Accelerators in Data Centers*,
https://doi.org/10.1007/978-3-319-92792-3

MIX
Papier aus verantwortungsvollen Quellen
Paper from responsible sources
FSC® C105338

If you have any concerns about our products,
you can contact us on
ProductSafety@springernature.com

In case Publisher is established outside the EU,
the EU authorized representative is:
Springer Nature Customer Service Center GmbH
Europaplatz 3, 69115 Heidelberg, Germany

Printed by Libri Plureos GmbH
in Hamburg, Germany